GARY MCKAY is a professional writer and has had seven books published. He was an officer in the Australian army and spent sixteen weeks in the clean-up operation of Darwin after Cyclone *Tracy*. In researching this book he interviewed over fifty of the cyclone's survivors—the 'riders of the storm' that wiped out Darwin.

GARY McKAY

TRACY

THE STORM THAT WIPED OUT DARWIN ON CHRISTMAS DAY 1974

ALLEN&UNWIN

First published in 2001

Allen & Unwin
83 Alexander Street
Crows Nest NSW 2065
Australia
Phone: (61 2) 8425 0100
Fax: (61 2) 9906 2218
Email: info@allenandunwin.com
Web: www.allenandunwin.com

National Library of Australia
Cataloguing-in-Publication entry:

McKay, Gary.
  Tracy: the storm that wiped out Darwin on Christmas Day 1974.

  Includes index.
  ISBN 1 86508 558 8.

  1. Cyclone Tracy, 1974. 2. Cyclones—Northern Territory—
  Darwin. I. Title

363.3492/0994295

Designed by Antart
Set in Times 11/13 pt by Midland Typesetters, Maryborough, Victoria
Printed by Griffin Press, South Australia
10 9 8 7 6 5 4 3 2 1

CONTENTS

PREFACE

WHEN I BEGAN researching this book, I had some idea of the task that lay ahead. When Tropical Cyclone *Tracy* destroyed Darwin on Christmas Day, 1974, I was a captain in the Australian Army. In early January 1975, I was sent to Darwin to assist in the massive clean-up operation of the devastated town. When I stepped from our chartered Qantas jet, I couldn't believe what I was seeing.

During the clean-up, I spoke to many people who had been through the cyclone. Others, understandably, weren't yet ready to talk about what they had experienced.

In 1971, in Townsville on the north Queensland coast, my wife and I, with our two dogs and a cat, had ridden out Tropical Cyclone *Althea*. I therefore thought I knew what a cyclone felt and sounded like. Having interviewed over fifty survivors of Cyclone *Tracy*, I now realise that my own cyclone experience could in no way be compared with theirs. The impact of that storm upon the town, and upon the psyche of all those who survived it, has been immense. The noise, the force of the wind and rain, and the bitter cold are deeply and indelibly etched in their memories.

I am indebted to all those people who opened up those memories and allowed me to include them in this book, as well as to those who provided valuable information but were unable to share their deepest feelings about the disaster.

This is a book about a devastating storm and those plucky, gutsy, courageous souls who rode it out. It is a story of miraculous escapes. Those same gale-force winds that bent steel girders at right angles to the ground left unharmed babies and young children who were exposed to the elements. The fortunes of topography and fate awaited the 43,000 inhabitants of Darwin on Christmas Eve, 1974, when *Tracy* was bearing down on them. Those whose lives were spared still carry the scars of their brush with death, but all have a zest for life that *Tracy* was unable to extinguish.

I dedicate this book to those who perished in that savage storm, and those who survived—the riders of the storm.

Gary McKay
June 2001

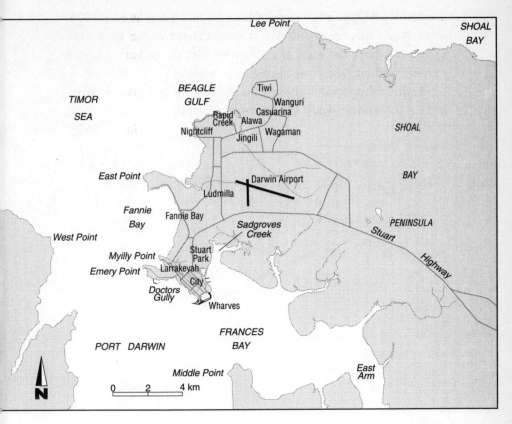

Darwin suburbs and surrounds, 1974.

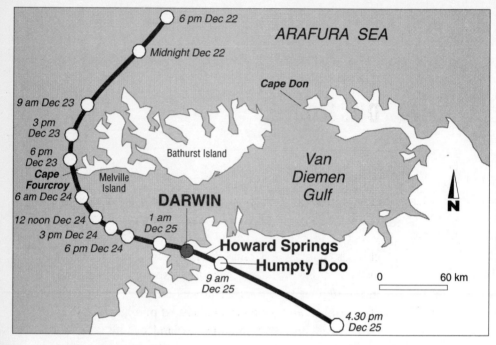

The path of Cyclone *Tracy*.

INTRODUCTION

IN DECEMBER 1974, Darwin, the capital of the Northern Territory, was a big country town, not yet a city. Admittedly, it was a boom town.

Originally settled in 1869 and called Palmerston, the town had been a backwater until the Second World War when its strategic value soared along with the expanding Japanese 'sphere of influence' in the region. When Australia was directly threatened, thousands of Australian and Allied troops poured into Darwin. The roads were improved and port facilities expanded, and general amenities soon followed. Then the Japanese bombed Darwin sixty-five times, destroying half of the buildings and scattering the population. After the war, many ex-servicemen decided to stay on in the town, and it began to grow when mineral deposits such as uranium were discovered in the 'Top End', as the Northern Territory is often referred to.

In 1974, Darwin wasn't a pretty place. It was a bit ramshackle in parts, with a mix of typical British-inspired colonial architecture, housing mainly Australian government administrative departments, and a mushrooming conglomeration of mostly high-set, fibrous cement-clad, corrugated iron-roofed houses. The town was described by one Territorian as 'not so much a concrete jungle, as a fibro bungle'.

The Northern Territory was administered by Canberra through a system of satellite offices in Darwin. Office workers—mainly Commonwealth public servants and private company representatives—were located in one of several streets in Darwin proper. Male workers, of which there was a preponderance in the population, went to work dressed in smart shorts, long socks and open-necked shirts. If single, they tended to live in hostel-style accommodation. Helen Auld, who now lives in Adelaide, described the Darwin of the mid-1970s as 'a man's town'; it lacked the normal social graces of a capital city, she said. Despite being unsophisticated, though, the town had character and the people were friendly. Many of the residents had no extended family in the town, and so the local populace came to rely on one another for close personal support. Long-term residents numbered only in the thousands among the approximately 43,000 residents who were living in Darwin in 1974. When I first visited what remained of the town in early January 1975, I wrote to my wife, 'This is Frontier-ville.' Darwin was the sort of place where people could come to begin a new life—a place to wipe the slate clean, with no questions asked.

Apart from the mainly Anglo-Saxon population, there were 6000 to 7000 Chinese residents, many of whom were descended from the Chinese labourers who came into the country in the 1860s and 1870s as part of the gold and railway building booms. Most were now self-employed traders and merchants. There was also a large Greek community that thrived on the entrepreneurial nature of the town and the Top End. Senator Grant Tambling, whose family have been residents for several generations, described the population as 'a potpourri of cultures'; others described it as 'a melting pot for the various racial groups'. It was a distinctly cosmopolitan town, with around fifty nations represented in the Darwin demography in a survey conducted in the early 1970s. The town had a distinctly Asian flavour—salty plums were the kids' currency in the schoolyard. Peter Coombe, who managed Hunter's Car Rental Company at the time of the 1974 tropical cyclone, described the town as having 'a distinctly oriental influence, where takeaway food was satays, not meat pies'. It was 'Somerset Maugham stuff': 'Life revolved around buffalo, barramundi and barbecues. Living in Darwin was like being on a long summer holiday.'

Recreation was self-driven and centred around the many social and sporting clubs that flourished due to the relatively young population. The bulk of people were in Darwin 'for a good time, not a long time'. The town catered for almost every imaginable sport, but the game that really held the imagination of Territorians was

Australian Rules football. Everyone played it or watched it, regardless of colour, race or creed.

Almost everyone interviewed for this book said that Darwin was a town where everybody knew everybody else. On Saturdays, people went into town to catch up with friends and share the local gossip. People didn't feel the need to lock their houses or cars. 'You left your keys in your car ignition so that you wouldn't lose them,' said Chris Collins, who worked for the Department of Agriculture at Berrimah. Friends usually dropped in without notice, and appointments were rare for business dealings unless at a very high level. Most people entertained with barbecues under their houses (raised to provide a breezeway for cooling and to protect against flooding in the wet season, known as the 'Wet') on concrete pads laid expressly for that purpose.

The largest piece of real estate in Darwin was (and still is) the airport. The airport is unique in Australia in that it is a composite Royal Australian Air Force (RAAF) and civilian facility. In addition to the domestic and international services, there was parking for fifty or so light aircraft, as many of the pastoral and mining companies used planes to overcome the enormous distances between locations in the Top End. The major user of the airport was the Department of Defence, which had proprietor rights and ran the airport security, maintenance and radar installations. The RAAF had several hundred personnel on base, along with about a hundred married quarters and normal base installation facilities capable of handling anything from a helicopter to a 747 Boeing airliner.

The Darwin Port and Harbour facilities were located at the southeastern end of the central business district. The port, though not totally protected, was the best shelter available in the area. It had two wharves capable of handling two normal coastal trader-type ships of around 220 metres on the outside and mainly 30-metre prawn trawlers on the inside. These were called the Stokes Hill Wharf and the Fort Hill Wharf. Another wharf, imaginatively called the Iron Ore Wharf, was specifically designed for loading iron ore. It extended further out into the bay to cater for the larger vessels that utilised that facility. Mooring buoys in the general harbour area provided anchorage points for other vessels not needing to use the wharves continuously. The Navy patrol boats berthed inside the Stokes Hill Wharf at their own berth. Around four hundred vessels per year berthed at the port.

The main part of town that accommodated trading, business and government offices was situated at the very end of the Stuart Highway on the south-eastern end of Darwin Peninsula. On the north-western end of the peninsula was the Australian Army base of Larrakeyah Barracks. The Defence Department also had a major communications installation south of the town at Coonawarra. This was a joint facility and contained sensitive communication and intelligence monitoring facilities, as well as vital defence links for the region.

Private dwellings that were not apartments or council flats were in two main suburban areas. The oldest and original suburbs were only minutes from the CBD, being located along the shoreline around the coast and facing into the many bays that formed part of the Darwin Peninsula. The bulk of these homes were high-set, with beautifully lush and well-established gardens of palms, frangipani, bougainvillea, and tropical plants and bushes. The other main area of population was the northern suburbs, beyond the airport, which contained mainly government-built houses for their workers. They were a mix of mainly high-set fibrous cement-clad houses on piers, and brick, low-set Housing Commission homes.

The only major road link to Darwin is the Stuart Highway, commonly referred to as 'The Track'. A bitumen two-lane road, it was frequently cut by flooding during the 'Wet'. In 1974 there was, and still is, no rail link to Darwin. The town had a local rail infrastructure that serviced the local industrial and warehousing areas, but there was no passenger rail network within the Darwin environs.

The tropical climate was easy to endure during the dry season (the 'Dry') from May to October, with warm sunny days, relatively low humidity and cooling sea breezes at night. It was at this time that Darwin was inundated with tourists from down south. One resident said that 15 degrees Celsius was considered 'freezing'. The wet season, however, was enervating, with high, stifling heat and soaring—almost choking—humidity. When the daily rainstorms dumped hundreds of millilitres of rain in less than a few hours, minor flooding in and around Darwin was routine. Sometimes the town could be cut off by road for weeks at a time. Every afternoon, almost like clockwork, tropical rain and thunderstorms accompanied by stunning electrical storms would pour down on the residents, providing some welcome relief from the heat. Sometimes the electrical storms could provide some scary moments. Two weekends before Christmas 1974, Marcia and Rowan Charrington were in their backyard talking to a neighbour and, in Rowan's words, 'He got zapped off my chain wire fence, and a kitten he was holding in his arms went limp. The cat stayed like that for several days but managed to survive.'

The 'Wet' is also tropical cyclone time. These tropical revolving storms breed at an indeterminate rate, and the region averages three or four of them per season. The last time Darwin had been visited directly by a tropical cyclone was in 1937, when damage was estimated at 50,000 pounds ($100,000) and five people were killed. The only other time the city had been badly damaged by a cyclone was in early January 1897 when, according to the Bureau of Meteorology, a 'disastrous hurricane and phenomenal rainfall' almost completely destroyed the fledgling town. Trees were uprooted for 70 kilometres around Point Charles lighthouse on the western entrance to Port Darwin. There were many casualties at sea, and a schooner was driven so far up the beach that refloating was impossible. Since 1937, the four tropical cyclones that had passed through the Top End had had only minimal effects, with only minor flooding, some trees uprooted and moderate damage to buildings on Bathurst and Melville islands. Tropical Cyclone *Selma* came close three weeks before *Tracy*, with winds of 90 kilometres per hour on Bathurst Island, but no significant winds were recorded in Darwin.

Against this backdrop of a laid-back, bustling, tropical town with the festive holiday season almost upon the inhabitants, Tropical Cyclone *Tracy* was destined to call. Within the time it took for *Tracy* to travel from its birthplace some 700 kilometres distant, move down through the Arafura and Timor seas to the shores of the Northern Territory, then smash into Darwin, the lives of these happy-go-lucky Territorians would be changed forever.

A STORM IS BORN

1

> **"** She's a goer. **"**
>
> — GEOFF CRANE,
> SENIOR METEOROLOGIST,
> DARWIN BUREAU

1 THE GATHERING STORM

THE CREW ON board the trawler *Clipper Bird* were racing against time to reach Darwin for Christmas Eve. It was Friday, 20 December 1974. The weekend would find them still north of the Gulf of Carpentaria. The sea, a tepid 29 to 30 degrees Celsius, was normal with only a moderate swell.

The trawler was one of a fleet of six operated by a joint Australian–Japanese prawning venture. It was returning from a refit in Newcastle, on the New South Wales coast north of Sydney, in preparation for the banana-prawn season that would begin in January. It had a skipper and seven crew, including an engineer, first mate and bosun. One of the crewmen was eighteen-year-old Rob Perkins, a cadet with the joint venture company. Two others were university students who were working as relieving deckhands during their university holidays. Rob recalled seeing very little of the deckhands as the boat steamed towards its destination. They weren't used to the pitching conditions, he said, and kept 'a low profile'. The boat was 26 metres in length, weighed 240 tonnes and was powered by two large Caterpillar diesel motors. It was a typical prawn trawler with a large rear working deck and two side arms for trawling. The *Clipper Bird* was empty and making the best possible speed to be back in Darwin to celebrate Christmas. As Rob Perkins later observed dryly, 'That was a big mistake.'

Earlier in December, far to the north of Australia above the frozen expanses of central Siberia, the first ingredients for a storm that would soon wreak havoc half a world away had been assembled. A continuous column of dense, freezing air, chilled by one of the coldest northern winters in a decade, had formed above the frozen plateau. The weight of the air in the column had caused it to sink heavily to ground level. More cold air then rushed in above to replace the sinking air, pushing the column out from underneath and across the vast frozen Siberian steppe, over the Sea of Okhotsk and towards Japan. Then it commenced its journey south-west into the western Pacific.

Meanwhile, over northern Australia, the land and air were being cooked by the hot summer sun. The heated air rose, forming the zone of low pressure that the locals in Darwin call the 'monsoon trough'. The hotter it became, the more the air rose, and the more replacement air was sucked into the trough.

The once cold Siberian air was now being pulled towards the monsoon trough. As it travelled down past Borneo, over the Java Sea and across the Equator, it was warmed by the tepid seas it crossed. The warmer it became, the more moisture it picked up. By 20 December, the air mass was dense with moisture—and becoming even warmer and wetter as it approached the Top End of Australia.

When it hit Darwin's monsoon trough, the warm moist air began to rise and circle, the moisture condensing to form heavy clouds. And as the air condensed it gave off heat, so the air became hotter and continued to rise even faster. More and more air rushed in at ground level to replace the rising column of air. Strong winds whipped away the hot air at the top in order to suck the air coming in at the bottom up through the column. If this didn't happen, the warm, but now dry air would have dropped back down the column and snuffed out the storm, just like a candle. But in Darwin in late 1974, it *did* happen. All the ingredients were now in place for the mass of air to become unstable and develop into one of nature's most violent events—an intense tropical cyclone.

The word 'cyclone' comes from the Greek *kyklon*, meaning a wheel. A tropical cyclone is a particularly intense revolving storm that forms in the tropics in the Southern Hemisphere and the Bay of Bengal. In the Caribbean, the Gulf of Mexico and the Pacific coast of Mexico, the same storm is called a hurricane, named after *harucan*, a Central American God of Evil. In the western Pacific it is called a typhoon, from the Cantonese *t'ai fung*, meaning 'great wind'.

There are several essential ingredients required for a tropical cyclone—an intense low-pressure system—to form. First and foremost, warm seas are needed to make the air warm and moist. A cold sea only provides about the same amount of moisture for the air as dry land. The ideal breeding ground for a tropical cyclone is seas warmer than 26.5 degrees Celsius, like the Arafura and Timor seas near Darwin in the summer season.

Once the warm, moist air starts to rise, it needs to become even warmer to keep on rising through the atmosphere. As the air rises, condensation occurs and the moisture changes from water vapour to water droplets. Huge clouds then start to form. Heat is released and the air becomes warmer and rises even faster. More and more air rushes in at ground level to replace the rising column of air. Six to 10 kilometres up, the temperature of the rising air can be up to 8 degrees Celsius warmer than the surrounding air—more than enough to keep the spiral updraft rising. The air rushing in causes gale force winds, and torrential rain and storm conditions can result. The area in the centre of the updraft, which has light winds, no rain and little cloud, is called the 'eye'. It is eerily calm compared to the storm raging around it.

In the upper atmosphere, some 12 to 15 kilometres above the ground, the column of rising air must be ventilated or blown away so that the air under it can take its place. When all this happens at the same time and place, we will have a tropical revolving storm— a tropical cyclone. If something is missing, then the tropical cyclone won't develop. It might form, but then quickly fizzle out. Ninety-nine per cent of the white fluffy blobs of cloud that appear in the tropics every day in satellite images on our television screens just fade out after a couple of days because one of the ingredients is missing.

Hundreds of kilometres above, the US Environmental Satellite Services Administration weather satellite, ESSA 8, was in orbit around the earth, taking infrared images of the northern Australian region. As it hurtled past, it transmitted data earthwards to several weather stations in Western Australia and Darwin. ESSA 8 was one of three weather satellites that provided images for the weathermen, who called themselves 'mets'. The weather satellites at the time were comparatively crude, which made it difficult for the mets to accurately forecast weather patterns. Besides which, they watched over a huge area that was the birthplace of at least three, and sometimes as many as seven, tropical cyclones each cyclone season.

At 9.55 a.m. on the Friday before Christmas 1974, a large cloud mass centred in the Arafura Sea about 740 kilometres north-east of Darwin was being tracked by Weather Bureau regional director Ray Wilkie and senior meteorologist Geoff Crane in Darwin. Covering several hundred square kilometres, the low-pressure system, which was dumping rain over the warm Arafura Sea, didn't look unusual for the time of year. What they were looking for were any tell-tale signs that it was developing into something that was potentially far more lethal—a tropical cyclone.

Over the next twelve hours, the mass of cold air that had begun its journey over central Siberia would be drawn down into the area monitored by the Darwin mets at an ever-increasing rate. They knew that some of the ingredients were already in place to produce a violent tropical storm. For the moment, however, the weather system was still too remote to be tracked by their radar, which had a range of only 400 kilometres.

The first tropical cyclone of the season had bypassed Darwin only three weeks earlier. Tropical Cyclone *Selma* had formed about 350 kilometres to the north-west of Darwin on 1 December. For the next twenty-four hours it had moved directly towards Darwin, and by 10.00 a.m. on 2 December it had been about 300 kilometres almost due west of the town. By 10.00 a.m. the next day, it was less than 100 kilometres west, but then it had inexplicably done an abrupt

U-turn, first turning north across the west of Bathurst Island and then west. It had continued moving west until it weakened on 9 December some 1100 kilometres west of Darwin. For days, the mets had been sending out storm warnings advising the people of Darwin about *Selma*'s movements, but the storm had just fizzled out. Every other tropical cyclone that had formed in the area in the last couple of decades had done the same.

Like many people in Darwin that Saturday morning, 21 December, Elspeth Harvey went shopping. Christmas was only a few days away and she had decided to buy a new Christmas tree. Caught up in the holiday mood of the town, she also splurged on new decorations, tinsel and coloured lights. She was delighted with the tree and happily spent hours decorating it back at her home in Fannie Bay.

Another shopper, housewife and mother Shirley Gwynne, from the northern suburb of Wagaman, was out buying material with which to make doll's clothes for her two-and-a-half-year-old daughter, Tahnee. Her other major purchase for the day was a turkey, which she intended to cook on Christmas Eve for lunch the next day with friends.

In the northern suburbs, twenty-three-year-old Ann Gray was working her normal Saturday morning shift at Coles' store in the Casuarina Square shopping complex. Ann and her husband Peter had been married for only a year or so and were relatively new to Darwin.

By midday, the mets at Darwin's Tropical Cyclone Warning Centre, located on the eighth floor of the MLC Building in town, had evidence of a gathering storm. The duty officer was convinced that he had a potential tropical cyclone on his patch. The latest

ESSA 8 satellite photograph showed evidence of a circulation centre near latitude 9 degrees south, longitude 132 degrees east. The tropical storm had gathered enough energy to create a definite spiralling updraft, and the upper winds were ventilating the column, drawing more warm air into its spiral. Until the storm started to move, it wouldn't be given a name—but that time didn't look far off.

At 4.00 p.m. Central Standard Time (CST), the met duty officer classified the disturbance and issued the first tropical cyclone alert, describing the storm as a tropical low that could develop into a tropical cyclone.

The message went out to the broadcasting media, other meteorological offices and selected authorities. It prepared those who received it to take appropriate action if a tropical cyclone warning was issued subsequently. The alert would be reviewed every three hours until it was either cancelled or replaced with a tropical cyclone warning. A tropical cyclone warning would be issued as soon as the mets anticipated there might be gale-force winds on any part of the coast within twenty-four hours. The general public would be alerted through announcements on radio and television. The warning would be preceded by a siren that had been developed a few years previously when the tropical cyclone warning announcing the imminent arrival of Cyclone *Ada* on the Queensland coast in 1970 had been ignored.

For this first alert, the Tropical Cyclone Warning Centre wasn't even looking at Darwin. The warning was aimed at the small, isolated communities on the Top End's north coast—Croker and Goulburn islands, Maningrida and along the Arnhem Land coast.

Also notified were the Civil Defence, the police, the navy and army, the harbour master, the Weather Service Office at the airport, coastal radio stations, and the Postmaster General's outpost radio, which serviced remote communities. It also went to the newly formed National Disasters Organisation (NDO) in Canberra. If the alert was later upgraded to a warning, the list would be expanded to include government agencies, the power station, the hospital, Darwin City Council and the large commercial sea-traders, such as Burns Philp & Co., that operated in the islands. In the event that

Darwin had a full-fledged tropical cyclone bearing down on it, the mayor, the town clerk, the controller of civil defence and other community heads would also be notified. As many people as possible would have to be told to prepare for an emergency.

A second weather satellite infrared picture received that evening from the US National Oceanic and Atmospheric Administration weather satellite, NOAA 4, showed that the low pressure had developed further. The spiralling cloud could clearly be seen. The mets' suspicions were confirmed. When the duty met looked across the large plotting table at his colleague and said, 'She's a goer,' he got an affirmative nod in return. The cloud centre had moved only half a degree south and a degree to the east, but a tropical cyclone was born.

The tropical revolving storm was officially classified as a tropical cyclone at 10.00 p.m. on 21 December and given the name *Tracy*. The operations room swung into action.

2 THE LULL BEFORE THE STORM

BY MIDNIGHT ON Saturday, Tropical Cyclone *Tracy* was 250 kilometres from Darwin, close enough to be tracked by the weather radar. The radar was located out at the airport and away from the interference of the town's buildings and topography. It was capable of being powered by the airport's emergency power generator in a storm.

Getting a radar picture from the radar hut out at the airport to the mets at the office in town was a matter either of tracing the radar image off the screen by hand, or photographing the screen with a black-and-white Polaroid camera and then faxing the image to the mets.

From 3.20 a.m. on Sunday, 22 December, the radar operator stationed at the airport took Polaroid exposures about every thirty minutes. This procedure was followed right up until the radar lost power early on 25 December. The first radar images showed an area of converging spiral rain-bands between 80 and 240 kilometres to the north.

Tropical Cyclone alert number 2 was issued at 10.00 a.m.:

```
TROPICAL CYCLONE TRACY WAS CENTRED 000 KILOMETRES NORTH
NORTHEAST OF CAPE DON AT 9.00 P.M. AND MOVING WEST-
SOUTHWEST AT 12 KILOMETRES PER HOUR. GALES ARE NOT
EXPECTED ON THE COAST WITHIN 24 HOURS, BUT COMMUNITIES
BETWEEN GOULBURN ISLAND AND BATHURST ISLAND ARE ALERTED.
```

The differences between the alerts at this stage were minimal because *Tracy* was dawdling. In the time between the first two alerts she had only travelled about 60 kilometres. The eye of *Tracy* wasn't apparent until another reading later in the afternoon. In 1974 the cyclone positioning in warnings was relevant to the eye, or centre, of the tropical cyclone. Today, warnings advise of the position of the leading edge of the destructive winds, since that is where the immediate danger lies. Consequently, tropical cyclone warnings now concentrate on when communities can first expect gale-force winds to hit.

At 1.30 p.m., the radar identified a partial eye wall echo, indicating that the eye measured about 25 kilometres in diameter. From this time onwards, the centre of *Tracy* could be tracked with much greater accuracy as she meandered sluggishly in a south-westerly direction across the Arafura Sea and slightly away from the Darwin coastal region.

The storm was gathering strength, but only slowly. There were no indications that it would do anything other than continue travelling in its present direction. The people most at risk were the indigenous inhabitants of Melville and Bathurst islands. If *Tracy* continued moving south-west, it looked like she would clip the western side of the islands. The islanders had seen these storms before; they knew they would get heavy deluges of rain and that the wind would uproot trees, but it would then go away. There would be no fishing for a few days, but they figured that would be about the extent of it. As *Tracy* moved within twenty-four hours' striking distance of the islands, the alerts were upgraded to warnings.

Up at the Paspaley pearl farm at the southern end of Port Essington (an inlet on Coburg Peninsula) about 180 kilometres north-east of Darwin, a team of men from the Paspaley Pearling Company was working on the pearl racks. The work boat *Paspaley Pearl* was anchored in Barrow Bay just behind the abandoned settlement of Victoria and provided the working platform above the oyster racks. The trip up from Darwin had been a routine supply run that would also enable some essential work to be done before the holiday break. The skipper of the boat was twenty-three-year-old Nick Paspaley. Nick was keen to get the work done quickly. His wife in Darwin was due to have their first baby on Christmas Day. As Nick recalled, 'I had promised my wife that I would be home for Christmas and for the birth. I was planning to be home on the twenty-third. My wife didn't want me to go because the birth was only a week off, but I said to her, "It's just a couple of days. I guarantee I'll be back in time. I wouldn't miss it for the world."'

But now *Tracy* had developed and was lurking out to the west just beyond Barrow Bay. 'The weather looked a bit too bad to leave,' said Nick. Wisely, he stayed where he was, and he and his crew continued working on the pearl farm.

Nick had designed the boat specifically for pearl farm work. It was a modified version of a Japanese stern-trawling prawn trawler. Nick thought the *Paspaley Pearl* was 'a good sea boat, with a good sea hull'. Thirty metres long, it had a shallow draught of 2 metres and drew only a metre on the bow. The 80-tonne craft was driven by a Caterpillar 353 diesel power plant pushing out about 450 horsepower. In Nick's opinion, 'It wasn't the best sea-going engine, but it was okay in that boat. The main thing was that she was reliable.'

Even on the pearl farm in Port Essington, where normally it was pretty calm, the seas were reasonably chopped about. Nick had been monitoring the tropical cyclone alerts and knew that the seas outside would be very rough. Every couple of hours, he would go into the wheelhouse and check the ship's radar and listen to the maritime radio. The weather reports of where the tropical cyclone was heading indicated that it was basically going west-southwest past Port Essington. The options were very limited, Nick recalled. 'There wasn't any chance of going out in front of the cyclone. So we were stuck there . . . but I was very keen to get home for Christmas.'

Tropical cyclones don't travel in straight lines. They are basically a revolving mass of rain and wind. The path they follow is known as a trochoidal track. Their movement is much like the rear wheel of a circus clown's bicycle, which appears to go up and down while still travelling forward. The cyclone's forward speed is never consistent—the revolving storm's advance speeds up and slows down according to the way it goes through the trochoidal motions. Therefore the forward speed is always an approximation taken over several hours.

The mets in Darwin knew that pressure systems and wind could steer a cyclone and change its course, but at that time they knew little about the wind and pressure systems that operated in the vast area of the Arafura Sea and towards Indonesia. Consequently, they didn't know if *Tracy* would continue moving in her present direction or change course.

In Darwin on that Sunday afternoon, people were drifting back home after having spent the day fishing or sailing or otherwise relaxing. It was typical wet season weather, with rain activity, heavy humidity and high temperatures. The Weather Bureau in town had just received the latest radar image from the airport. *Tracy* was now shown with a fully developed eye. The mets were pleased in one respect, because the tropical cyclone would now be a lot easier to track. But the proximity of the cyclone's centre and the strength of the rain-band surrounding the revolving storm meant the people of Darwin would have to be warned.

Thirteen-year-old Nat Guinane was lying on the floor of her parents' flat in Nightcliff watching her favourite television program, *Countdown* on ABC television. The the pop music show was interrupted by an announcement that a tropical cyclone was in the region.

```
TOP PRIORITY. FLASH. CYCLONE WARNING NO. 1 ISSUED BY
THE DARWIN TROPICAL CYCLONE WARNING CENTRE AT 4.15 P.M.
CST, 22 DECEMBER 1974.
AT 3.00 P.M. TROPICAL CYCLONE TRACY WAS CENTRED
```

```
80 KILOMETRES NORTH OF SNAKE BAY AND MOVING
WEST-SOUTHWEST AT 9 KILOMETRES PER HOUR.
THE CENTRE IS EXPECTED TO BE NEAR BATHURST ISLAND AT
3.00 A.M. TOMORROW.
DESTRUCTIVE WINDS OF 80 KILOMETRES PER HOUR WITH GUSTS
TO 120 KILOMETRES PER HOUR ARE EXPECTED ON BATHURST AND
MELVILLE ISLANDS TONIGHT AND EARLY TOMORROW. FLOOD RAINS
ARE LIKELY OVER BATHURST AND MELVILLE ISLANDS.
ABNORMALLY HIGH TIDES MAY OCCUR ALONG THE COAST BETWEEN
GARDEN POINT AND CAPE DON.
```

When Nat sat down with her parents and brother for their evening meal, no one commented on the news flash.

Snake Bay is a large inlet on the northern side of Melville Island and has a township of several hundred people. Garden Point is a major settlement on the western side of the island opposite nearby Bathurst Island. Bathurst and Melville islands are situated about 50 kilometres directly north of Darwin and form an almost geographic chain that encompasses the waters of the Van Diemen Gulf. The island group covers approximately 170 kilometres from east to west and about 70 kilometres from north to south. Both islands are Aboriginal reserves, and the inhabitants follow a traditional way of life by hunting, gathering, collecting and fishing. Bathurst is the smaller of the two islands and is separated to the west of Melville by a narrow strip of water. Thirty kilometres to the east of Melville Island, on the tip of the mainland Coburg Peninsula, is Cape Don, a natural maritime checkpoint for all sea-going craft passing through these waters. On the westernmost end of Bathurst Island is Cape Fourcroy, the site of a Department of Science automatic weather station which recorded and transmitted weather data direct to the Meteorological Bureau offices in Darwin. Its main task was to measure and record wind strengths and rainfall.

Dianne Ferguson, the mother of two boys aged under five, worked part-time at the Darwin RAAF base wet canteen. The airmen relaxing in the club that afternoon were talking about the storm and whether it might affect air operations and disrupt the Christmas stand-down. Later that night, Dianne and her husband Bob went to a friend's house for dinner and the subject of the storm came up again. The dinner guests discussed at length whether the storm would reach Darwin. Dianne recalled that the hostess thought it would, and 'bets were placed on the likelihood of it coming in'.

On the other side of the airport, Bill Gough, a maintenance foreman and engineer with the regional carrier Con Air, was vitally interested in the storm because it would affect his crew and their workload close to Christmas. However, like many others, he remembered how Cyclone *Selma* had come close to Darwin and then veered away to die as a rain depression down south.

Rowan Charrington, who lived in the northern suburb of Wanguri, had taken the precaution of bracing and tying down the young mahogany trees in his yard when *Selma* approached the town. There had been a lot of rain, and many trees had blown down, even though *Selma* had stayed out to sea. The ground had been saturated to the point where root systems lost their grip on the soil.

By Sunday night, *Tracy* was about 100 kilometres north of Melville Island, moving at between 6 and 12 kilometres per hour. Her track was still generally in a south-westerly direction.

At first light on Monday, 23 December, the storm was within 10 kilometres of Melville Island's northernmost point. The gale-force winds were bending palm trees and stripping the foliage off the fronds. It was pouring down in bucket-loads. The Aboriginal and islander people battened down for a 'big blow'.

In Darwin, people were heading off for a day at work, their last full day before breaking for the Christmas holiday. Around 9000 people on Christmas leave had left town over the weekend by air or road. Tomorrow was Christmas Eve, and in Darwin that meant 'party time'. The larger hotels and clubs were all planning activities, and thousands of Darwinians were planning to socialise with friends at home over a barbecue and drinks.

Around lunchtime, as if it was a portent of things to come, an earth tremor shook Darwin. Julie Tammo, a housewife and mother of two young boys, was at home in Packard Street, Larrakeyah, making lunch for her children when she felt the ground shake. The ground around Larrakeyah, on the northern point of the peninsula, is rocky—'tough dirt'. Julie felt her old government-built house shake and move. It was a typical 1950s' house on steel piers with wooden floors and fibrous (fibro) cement walls punctuated with banks of glass and metal louvres. She grabbed her two sons, Rohan, aged four, and two-year-old Brendon and sprinted downstairs and out into the yard. Her first thought was to call her firefighter husband, Heikki, who was at work. However, when there were no more tremors, she calmed down and then went back upstairs with the boys.

In town, most office workers were out of their buildings grabbing a bite to eat or doing last-minute Christmas shopping when the tremor struck. Marcia Charrington remembers it as being 'quite severe'. The high-rise building where she worked sustained a long crack in the end wall, from the top of the building all the way down to near the lower floors. When work resumed for the afternoon, everyone was talking about the tremor.

Out at Berrimah, some 10 kilometres east of the town centre, twenty-three-year-old Department of Agriculture worker Chris Collins was working at the department's research farm when he felt the ground shake. He didn't pay it too much attention, but he remembered it well enough to recall the incident twenty-five years later.

The tropical cyclone was still drifting west-southwest, and not much would change in the next twelve hours. A radar photograph taken at 4.45 p.m. when the centre was 145 kilometres north-west of the airport radar showed a far more mature and well-organised tropical cyclone than the previous photograph. The eye was now fully developed with a diameter of about 13 kilometres, and the spiral rain-bands comprising the rain-shield showed a well-organised structure.

A double concentric eye wall structure was also indicative of a well-developed tropical cyclone. The mean wind-speed of the system was now around 100 kilometres per hour. The ocean around the storm was being whipped into a foaming fury. The rain belting into the sea was now almost horizontal with the ocean surface, and it was impossible to see more than 15 metres ahead through the lashing rain and spray. *Tracy* was coming of age.

By mid-afternoon over on Melville Island, no one was outside. Families in Snake Bay and Garden Point huddled together for warmth as the wind and rain brought cool conditions to the tropical islands. It was difficult to walk or run in the rain, which stung the face, arms and legs. The mud on the ground sprayed everywhere like red spray paint. The sky had darkened, though it was still several hours until sunset. Dogs sheltered miserably under the huts facing the beaches and tried to find some relief from the sand whipping past their noses. Some empty kerosene tins bounced down the track between several houses and punctuated the cry of the wind as it tore at the fibro shacks. The rain was torrential. At 6.00 p.m., *Tracy* clipped the western end of Melville Island with gale-force winds and severely damaged the Snake Bay and Garden Point aboriginal settlements. Buildings were smashed, roofs were torn off dwellings and most trees were stripped of large foliage. The automatic weather station pluviograph (rain gauge) recorded 220 millimetres of rain at Garden Point in twenty-four hours.

Between 4.00 and 7.00 p.m. on Monday night, *Tracy* had been stationary—but then she started moving again. The eye appeared to move west, south-west and south-east over a period of three hours. At this time, the eye was fully developed and measured around 13 kilometres in diameter. Forward progress was less than 10 kilometres during that time. The forward speed was difficult for the weathermen to predict, as the storm was moving so erratically and slowly. The average speed was around 5 kilometres per hour. The 45-kilometre stretch of coast between Snake Bay in the east on Melville Island and Garden Point was warned that they could expect abnormally high tides. The tropical cyclone eye was now about 40 kilometres from Garden Point. Between 7.00 p.m. and midnight, radar pictures continued to be transmitted showing *Tracy* drifting west-southwest. The tropical cyclone had now entered the Timor Sea. Although it was raining violently on Melville and Bathurst islands, Darwin appeared not to be in any immediate danger.

3 IMPENDING MENACE

BY MONDAY EVENING, the eye of the tropical cyclone had passed Port Essington and Nick Paspaley had decided to have a shot at getting the *Paspaley Pearl* back to Darwin. He wanted to keep the promise he had made to his wife and be back in Darwin in time for the birth of their child.

Nick had been following the course of the tropical cyclone closely through the reports on the coastal radio. He decided to follow the eye of the storm, as he figured the cyclone would move away and head down the coast of Western Australia, which was what these storms usually did. Nick's plan was simple: he would head out of Port Essington and along past Cape Don, then go south-west through the Van Diemen Gulf and the Vernon Islands channels to Darwin. But to head out of Port Essington, he had to head out to the north-west, and to do that he had to head straight into the huge swells that had been built up by the tropical cyclone. He had to get the *Paspaley Pearl* far enough out from the headland so that he could safely turn the boat south. If he turned too soon, he would risk being broached in the huge swells that were pushing across the sea.

By the time Nick got far enough out to make the turn safely, the conditions weren't as intense and he was able to cut across the swell in relative safety. However, what usually took the *Paspaley Pearl* two hours had taken five or six, and it was now night, with low cloud and no moon. Nick described what happened as he started his run south:

 I took off about twelve hours behind the eye and followed it down to Darwin. There was coastal radio, but at the time it wasn't very clear where the cyclone was. I had a shocking trip. I actually thought of turning back a few times. From about Port Essington across to Cape Don, we got the nor'west winds. It was rough with big seas, but it wasn't quite as bad as heading straight into it. But then I had a problem when I had to turn south to turn down into Van Diemen Gulf—I was surfing down these enormous waves in a [23-metre] boat! We were surfing down the face of the waves, and I was in the wheelhouse doing two things at once. I had about six in the crew, and I had a few passengers who were friends and farm crew, and people like that. My cousin Nick Dondas [later to become a Member for the Northern Territory] was one of them. It was an exciting ride and I had to run out of the wheelhouse at every wave to see what the stern was doing. So, I would steer down the wave and go across the wave. Or the wave would be coming and I would check the course, and then I would run out the back and see what the stern was doing. The boat handled the seas not too badly, but I wouldn't do it again!

Nick had no idea of the size of the swells they were surfing down in his 23-metre boat. He recalled that they were very long, but because it was dark he couldn't judge the distance between the swells. But he knew they were big. The boat was careering along with its running lights on, and Nick had the boat's spotlights on so that he could see the seas. As Nick recalled:

I couldn't have managed it without the spotlights. I had one hand on the throttle and one hand on the steering wheel. I kept cutting back on revs and speeding up again to keep my nose out of the trough. It was pretty rough. My brother-in-law was on board and he was so seasick he was just lying on the back deck, which was awash with waves. I had to go down and tie him up to the gunwale, otherwise he was going to wash over the side. Despite several of the crew being a bit crook, there was no terrifying anxiety. They weren't particularly worried about the cyclone, because none of them had gone through one. We'd never been hit by a bad one. I've ridden out a few cyclones, but they were 70 knots, 100 kilometres of wind—but none of us had ever experienced a devastating cyclone. It was just another storm, a bit exciting because it was a cyclone, but there was no fear of it. They didn't like the rough weather, and every now and then they would come over and say, 'Is everything all right?' and I would say, 'Yeah.' But I had the watch to myself, because there was no one else.

It was a slow trip for the *Paspaley Pearl* except when the sea was behind them, and then it was quite fast for a couple of hours. It usually took thirteen or fourteen hours to travel down from the pearl farm at Port Essington to Darwin, but this trip took them about twenty-four hours. Nick said it was an experience of a lifetime, but one he didn't wish to repeat. Nevertheless, when I was talking with Nick Paspaley, I sensed that the experience gave him a tremendous amount of personal satisfaction. As he said, 'I felt that I was in control. It was a bit scary, though, because I wasn't sure the boat could do what it did. But it was one of those vessels with the entire superstructure forward and a large back working deck, so it handled it superbly.'

At midnight on the Monday, Cyclone *Tracy* stopped following the script. Like most other tropical cyclones, she was supposed to continue on a south-west track that almost parallels the coastline and scare people further south. Between midnight on the 23rd and 12.30 a.m. on Christmas Eve morning, the tropical cyclone moved parallel to the coastline, travelling quite slowly. At around 1.30 a.m., she almost stopped. The next radar image, taken at around 2.30 a.m., showed that she had started moving south-southwest again and was opposite Cape Fourcroy and about 20 kilometres out to sea. For the next two hours, she meandered in a small area south-west of Cape Fourcroy. It was almost as if she was loitering with intent and trying to decide which way to go.

The first direct evidence of the strength of the winds was at 3.30 a.m. when the Meteorological Bureau's automatic weather station at Cape Fourcroy recorded a mean wind of 100 kilometres per hour from the north-east and an air pressure of 993 millibars.

Three hours later, at first light on Christmas Eve, *Tracy* was about 30 kilometres out to sea off the south-western end of Cape Fourcroy on Bathurst Island. At 6.30 a.m., the mean wind observed at Cape Fourcroy was 110 kilometres per hour from the

north-northeast as the tropical cyclone moved slightly closer
in and rounded the western tip of Bathurst Island. At 9.30 a.m.,
the wind at Cape Fourcroy moved back to the north-west and
increased to a mean of 120 kilometres per hour as the centre of the
cyclone, now about 15 kilometres from the station, changed
course. For the past three days, this tropical revolving storm had
been slowly intensifying and moving in a south-westerly direction over a track of about 250 kilometres. When people turned on
their radios on Christmas Eve morning, they heard the now-
familiar siren noise followed by the stentorian voice of the
announcer:

```
PRIORITY. CYCLONE WARNING NO. 14 ISSUED BY THE DARWIN
TROPICAL CYCLONE WARNING CENTRE AT 7.00 A.M. CST,
24 DECEMBER 1974.
AT 6.00 A.M. CST SEVERE TROPICAL CYCLONE TRACY WAS
CENTRED 120 KILOMETRES WEST-NORTHWEST OF DARWIN AND
MOVING SOUTH AT 4 KILOMETRES PER HOUR.
THE CENTRE IS EXPECTED TO BE 100 KILOMETRES WEST OF
DARWIN AT 6.00 P.M. CST TODAY.
VERY DESTRUCTIVE WINDS OF 120 KILOMETRES PER HOUR WITH
GUSTS TO 150 KILOMETRES PER HOUR ARE EXPECTED TO
CONTINUE ON BATHURST ISLAND TODAY.
```

The mets decided to check the upper winds at around 7.30 a.m.
They released a weather balloon, which was then tracked by the
radar. When tracking the balloon, the radar cannot take images of
the tropical cyclone, so for two hours there were no images of what
Tracy was doing. The next radar image at 9.15 a.m. showed a very
different pattern of movement. The tropical cyclone had done a
sharp left-hand turn and was now heading east. The change of
direction was totally unexpected.

Tracking a tropical cyclone was relatively simple; what was difficult was predicting where she was headed. Geoff Crane was on
duty in the operations room that morning and was studying the
radar photographs and other related data. The right-angle turn was
mystifying. He recalled:

> When we got the radar pictures back again, suddenly it wasn't south; it was—oops! It was east! We couldn't see any reason for it, so we watched it closely. Initially, we wondered whether it wasn't in fact another short-term perturbation in what might have been an otherwise continuous south-west movement. It took a few hours for us to be convinced that *Tracy* had changed direction and was coming in towards Darwin.

Predicting tropical cyclone tracks wasn't a science in 1974; it was more of a black art. It was done by extrapolation of present data and by looking at associated weather systems in the region. Geoff Crane and his boss, Ray Wilkie, were now faced with a dilemma. 'We had to make some serious decisions about what we were going to tell the citizens of Darwin.'

The mets looked for indications as to why *Tracy* had turned east. They checked weather systems from 500 to 1000 kilometres all around the tropical cyclone to determine where it might go next. They were looking for a significant synoptic feature, such as a big front coming up with big winds like the north-westerlies, that might try to blow *Tracy* off to the south-east. But they knew this didn't tend to happen at these latitudes. That sort of system with an accompanying trough happened further south, down near the Pilbara latitudes, not across the Top End. Troughs at Darwin's latitude of 10 to 11 degrees south just didn't happen. So all eyes were searching for irregularities in the cloud blob. They needed some sort of indicator that would steer the tropical cyclone in another direction, or an indication of why it had so dramatically changed course. But there were none.

Tropical Cyclone *Tracy* was now just 115 kilometres west-northwest of Darwin. Geoff Crane decided to wait for another hour or so to see what was happening. The radar images taken between 9.15 and 11.30 a.m. weren't what the mets wanted to see. The tropical cyclone was still loitering with intent, and actually moved back towards Bathurst Island for a brief period before again heading south-east—towards Darwin.

While the mets were waiting for another radar image between 10.00 and 10.30 a.m., Geoff Crane slipped out to do some last-minute shopping in town. At the bank, he ran into Tony Pickering, a fellow tennis player, and gave him the lowdown on the cyclone. As Tony recalled, 'We had had Cyclone *Selma* only a few weeks before and suffered ten days of cyclonic weather. Everyone was sick of talking about cyclones—and besides, Christmas was here. I went back to work!'

There were several senior mets on duty that morning. Jim Arthur, who is now the regional director of the Bureau in Darwin, was rostered to go to the ABC studios to do a live radio talk. The show was 'Country Hour' and it went to air about 12.30 p.m. Jim had to be at the studio about half an hour ahead of time.

Geoff Crane and Ray Wilkie had been looking at the tropical cyclone on the chart coming in from the east for several hours. They now had six separate tracking locations from radar since it had turned. Geoff Crane recalled:

> We hadn't actually changed our warnings, but we were getting very agitated about it. I had been talking to our other senior met, George, out at the radar and by about this stage I had seen enough. Somewhere around midday or just before, we agreed that it looked grim. Jim Arthur was scheduled to give his talk on the weather at around 12.10 or 12.15 p.m. So I had to ring up the ABC and pull Jim out of the studio where he had just walked in. I said to him, 'Tear up that weather story. Here's your new story.'

High above the water on the cliffs overlooking the sea, Rick Conlon was peering towards the tropical cyclone area. He had seen many tropical storms and had experienced a few close misses with tropical cyclones during his thirteen years in Darwin. He thought the blue sky looked somehow brooding—almost malevolent. Rick was a refrigeration mechanic by trade and a senior maintenance officer at the hospital. He spent the morning with the maintenance crew working frantically to prepare for the coming storm. The crew taped the large glass windows in the casualty and reception areas and prepared all of their emergency repair kits in case the worst came to pass.

Inside the hospital, customs officer Greg Huddy was collecting his wife, Barbara, and their four-day-old baby girl. They would be going home to Stuart Park, an inner suburb that overlooked Frances Bay. The Huddys now had three children, three-year-old Nicole, one-year-old Peter and now baby Suzanne. Barbara wanted to be home for Christmas Eve. The government contractors had painted their Customs house in Westralia Street while Barbara was in hospital, so all their clothes and personal belongings were in plastic bags in the linen press to keep them out of the way of the painters. Barbara recalled Greg saying that the tropical cyclone was headed towards Darwin. Greg had spoken to Ray Wilkie from the Weather Bureau, and he assured Barbara that they were better off in Stuart Park than in many other places, because the houses in Stuart Park were mostly in the lee of a hill.

Eddie Quong, a third-generation Darwinian, knew the storm was on its way, but he wasn't concerned. His mother had been through the last tropical cyclone to hit Darwin in 1937 and he figured it would be 'just like another nasty wet season monsoonal storm with a bit more wind'. There were more important things to think about, like spending Christmas Day with his extended family, including his son who was home from Adelaide for the university semester break. At home in Shepherd Street, almost in the town proper, Eddie's wife Greta was busy in the kitchen of their wooden high-set house preparing a feast for the next day.

At the Weather Bureau, Ray Wilkie had been on the telephone non-stop all morning. Among the many people he had spoken with was the Darwin Port harbour master, Carl Allridge. Carl had lived in Darwin for more than thirteen years and had been harbour master since 1966 when the previous harbour master had died in a drowning accident. He had a staff of about thirty men, including two assistant harbour masters and four pilots who guided the larger vessels into port and berthed them in the tricky port conditions. The depth of water at the berths varied from 12 to 13 metres and was at least 16 metres in the centre of the port. The unusual aspect of berthing at Port Darwin was the very large tidal range of almost 9 metres and a strong tidal flow that accompanied the movement of water in and out of the bay.

From Carl's office at the end of Stokes Hill Wharf, he had a view of the working wharves. The wharf precinct of the Stokes Hill Wharf and the Fort Hill Wharf catered for medium to large vessels and formed a perimeter of sorts for the inner berthing of smaller craft and the prawning fleet. The prawn trawlers were steel vessels of between 20 and 30 metres in length. In addition to several local ferries that serviced the local island groups, various government agencies, the postmaster general, the police, customs and oil companies also berthed a variety of craft within the port facility.

The Royal Australian Navy (RAN) also headquartered its fleet of four Attack Class patrol boats in Port Darwin. The HMAS *Advance*, HMAS *Arrow*, HMAS *Assail* and HMAS *Attack*, almost 33 metres in length and weighing 146 tonnes when fully loaded, acted as coastal surveillance and intercept vessels, monitoring the movement of foreign vessels, including fishing craft, in Australian waters, as well as assisting in Customs and Immigration operations and search and rescue. Normally, the patrol boats were berthed on the inside face of Stokes Hill Wharf.

After talking to the mets earlier, Carl Allridge had ordered several of the larger vessels out of the harbour when the tropical cyclone was showing signs of menacing the port. Two of those vessels, the *Raima Kathleen* and the *Lady Cynthia*, an oil rig tender, were now close to the tropical cyclone out at the entrance to Beagle Gulf, the body of water between Bathurst and Melville islands and the Darwin coast. Around midday, when it appeared that the tropical cyclone was heading for Darwin, all vessels remaining in port, including the Attack Class patrol boats, were ordered to leave the wharves and anchor in the harbour precinct or seek shelter elsewhere. There were at least twenty-seven vessels in the immediate vicinity of the wharf or inside the port area; another fifty vessels of various sizes were taken out to their cyclone moorings inside the harbour area but away from the port precinct. One of the prawn trawlers that was tied up on the outside of the Iron Ore Wharf was the *Clipper Bird*.

Rob Perkins, the cadet seaman on board the *Clipper Bird*, recalled that they had had the good fortune to get through the rough ocean and into the protection of the island groups before *Tracy* started heading towards the coast. Normally, the *Clipper Bird* would have tied up on the inside of the wharf, but there were a lot of vessels in port for Christmas and wharf space was at a premium. The crew had cleaned up the boat and made it ready to put to sea again if required. The crew then dispersed, the majority heading off to their local watering-hole. When Carl Allridge later gave the word for boats to be moved out of the harbour, the men had to be tracked down.

When Rob Perkins and his shipmates were called to the telephone and told they had to return to their trawler and put to sea, they bought a few cartons of beer before heading back to the port. The skipper of the *Clipper Bird*, an Englishman who had just been given his first command, would be severely tested by the angry seas during the next sixteen hours. The water outside the port was now very rough, the swell was lumpy and messy, and the wind was 'pretty blowy'. The plan was to put out into the harbour and anchor up into the wind.

The majority of the vessels asked to leave port did so, but approximately six vessels were crew-less and immobile. These craft were prawn trawlers that either had their engines out for refit or were undergoing other forms of maintenance. While that work was being done, their crews took the opportunity to take some leave and usually went interstate or on a holiday elsewhere. Carl Allridge's problem was to make those vessels fast and tie them securely to the inner wharves. The last thing he wanted was a 30-metre, 200-tonne craft loose in his harbour. His assistant, Colin Wood, was busy ringing around the trawler companies getting their boats moored and organising other craft to be secured. The pilot boats were moored at buoys in the inner harbour. Greg Huddy worked for the Customs Department and recalled that Ray Wilkie had rung. 'Ray advised us to get the Customs launch, a 25-footer, out of the water and into an air-raid shelter down near the wharf. The launch master, Bluey Walters, moved the launch.' They had no mooring buoy as they usually had their own wharf space, and so taking the launch out of the water was the next best option.

Police Sergeant John Woodcock was a station sergeant at the main police station in Darwin. Like other government agencies, he had been tracking the tropical cyclone and was well informed of what was occurring. 'But it was still being taken lightly,' he recalled. 'It was "just another blow". People were probably complacent.' That Christmas Eve a batch of new recruits had arrived and were standing around waiting to be allocated to their new shifts. John set them and other police officers to work preparing vehicles, checking spare tyres, arranging the emergency equipment and wet weather gear, and generally going through the routine for a civil emergency. As was normal procedure, an Incident Room had already been established in the Police Operations Centre. Officers from Police Communications were now in constant touch with the mets at the Bureau.

The suburb of Stuart Park directly abuts the Darwin CBD. It's a mix of strip shopping precinct and residential housing. Chris Kingston-Lee, a four-year resident of Darwin, owned a motorcycle shop along the main road and had ridden into town that morning on his own bike from the northern suburb of Rapid Creek. He had had to take a bit more care than usual, as it was wet on the roads. It wasn't raining heavily in the morning, but it was breezy. Chris was looking forward to finishing some recent sales and registering some motorcycles for their new owners. He had enjoyed a good trade this Christmas and his shop display had won a local competition for the best Christmas window display. Instead of reindeers, motorbikes pulled Santa's sleigh, and the display had drawn a lot of people to the shop. Things were good. Chris recalled that 'people were mentioning the tropical cyclone that was out near Bathurst Island'. But the general view among his customers was that it would turn away and head down the coast, as usual. 'They seemed more concerned that the storm would ruin Christmas,' he said. Chris was inclined to agree. He had seen Tropical Cyclone *Selma*, which he thought was no worse than a severe storm. His main concern was the wind blowing over a huge tree in his yard if the ground got really wet and the root system was weakened.

Over at the RAAF base in one of the married quarters, Josephine Foreman had been preparing for the storm. She had her hands full. Her son Martin, one of two children, had been born with spina bifida and was effectively a paraplegic. Josephine had to carry him everywhere. The RAAF had a system for warning residents on the base about impending storms. Every three hours, a truck drove around the married quarters with a large public address system on the back and warned of the impending tropical cyclone. After a while, though, the warnings just became 'background noise'. But Josephine was well prepared. She had assembled the family's camping gear into bags and organised spare clothing, torches, candles and, because she was a trained nurse, a very good first aid kit. She had spent the day cooking a large turkey for Christmas lunch. Her husband was at home on stand-down from the RAAF where he was employed as a sergeant technician. The family spent the day mostly under the high-set house where the kids were playing in a small wading pool. Josephine's husband liked a drink and he spent most of the day working his way through a carton of beer.

Out in the northern suburbs at Nightcliff, Lesley Mance was helping her sister-in-law get things ready for Christmas Day. Fifteen-year-old Lesley was in Darwin visiting her older brother John, who was a policeman on shift work. She and her fifty-two-year-old mother had arrived two days before, and high on Lesley's list of priorities was catching up with her one-year-old nephew. The house they were staying in was a government house owned by the Northern Territory Police Force. It was old by Darwin standards and had been built in the boom period immediately after the Second World War.

The suburbs of Rapid Creek and Nightcliff sit side by side on a small outcrop of land about 2 kilometres across and the same deep. Both suburbs front on to the ocean and are purely residential, with small clusters of shops in each suburb and a larger Casuarina shopping complex on the main arterial road to their south. Thirty-year-old Barbara Crane, whose husband Geoff was senior met at the Weather Bureau, was at home in Rapid Creek with their two children aged five and three. Barbara wasn't working full-time, but was heavily involved in organising pre-school activities. She recalled that it was raining, windy and fairly overcast in the morning. During the day she had received several telephone calls from Geoff, who was keeping her updated on what was happening. He sounded concerned. When he rang just before lunch, Barbara started to feel uneasy and apprehensive.

At 27 McKeddy Street in Jingili, just a few kilometres from where the Cranes lived, Karen Jurek was waiting patiently for a television antenna installer to arrive. Her husband Kevin, who was a foreman of diesel generation with the Department of Works, had bought a new colour television set for Christmas—despite the fact that Darwin didn't have colour TV at that time. Kevin was at work, and Karen was looking after their three very young children—Kim, Nicole and Megan—a bull terrier called Kaiser, a clutch of rabbits and a cat. The Jureks' home was a Housing Commission pre-cast concrete, three-bedroom dwelling made of brick. It had louvred windows, an iron roof and was low-set. They had been living in Darwin for about six years. Kevin's sister also lived in Darwin, and she had arrived earlier in the day to go shopping with Karen. However, Karen thought it was too wet to take the kids out—and besides, she was waiting for the antenna man. He was never to make the house call.

After lunch, most companies and government agencies around town stopped work for the day and Christmas parties started cranking up. At Larrakeyah Barracks, the base had a population of about three hundred Regular Army personnel. Captain Peter Kerntke was attending the barracks' all ranks Christmas party, but he wasn't having too great a time of it. He had one leg in plaster up to the knee from a previous good time he had had in the Officers' Mess playing indoor soccer. Peter and his wife Diane lived in an army married quarter on Nurses Walk, a road that runs parallel to the cliffs overlooking Doctors Gully. It's an ideal position from which to observe the fantastic electrical storms that fill the black skies over Beagle Gulf in the 'Wet'. The cliff edge was 100 metres from their high-set standard government house. While Peter was at the Christmas party, Diane was bent over her sewing machine making party dresses for her daughters aged nineteen months and six years. Peter was communications officer at army headquarters, as well as the officer commanding 125 Signal Squadron, so he had both tactical and strategic communications to maintain. He was also rostered as a duty officer in the operations centre in times of emergency.

By midday or shortly after, *Tracy* began to move more determinedly towards the Darwin coastline. It began to gather strength as it passed over a band of very warm water estimated to be about 31 degrees Celsius-plus in the area south of Melville Island. This warm body of water was very large and extended well into Beagle Gulf. A marked intensification of the tropical cyclone took place at this point. At 1.00 p.m., a ship, the *Raima Kathleen*, located about 30 kilometres east-southeast of the centre of the tropical cyclone, reported continuous heavy rain, a north-northeast wind of 95 kilometres per hour and a fall in the barometer reading of 4 millibars in thirty minutes. Another ship, the *Lady Cynthia*, was located about 75 kilometres north-west of the centre. The mets in Darwin expected the observations of the *Raima Kathleen* and reports from the weather station at Cape Fourcroy to indicate that pressure to the north-west of the centre was rising and the winds were moderating. Instead, the *Lady Cynthia* indicated that winds had *increased* from 30 kilometres per hour at 12.30 p.m. to 70 kilometres per hour at 3.30 p.m., with a fall in pressure of 4 millibars in this period. The storm was becoming significantly stronger. *Tracy* was now 110 kilometres north-west of Darwin and heading south-east. It didn't take an Einstein to figure out what lay in store.

TOP PRIORITY. FLASH CYCLONE WARNING NO. 16 ISSUED BY
THE DARWIN TROPICAL CYCLONE WARNING CENTRE AT
12.30 P.M. CST, 24 DECEMBER 1974.
AT 12 NOON CST SEVERE TROPICAL CYCLONE TRACY WAS
CENTRED 110 KILOMETRES WEST-NORTHWEST OF DARWIN AND IS
NOW MOVING SLOWLY SOUTH-EAST CLOSER TO DARWIN.
THE CENTRE IS EXPECTED TO CROSS THE COAST BETWEEN GROSE
ISLAND AND THE VERNONS TOMORROW MORNING.
VERY DESTRUCTIVE WINDS OF 120 KILOMETRES PER HOUR WITH
GUSTS TO 150 KILOMETRES PER HOUR HAVE BEEN REPORTED
NEAR THE CENTRE AND ARE EXPECTED IN THE DARWIN AREA
TONIGHT AND TOMORROW.

2

THE SOUND AND THE FURY

❝ It was like a thousand express trains coming at us. **❞**

— CHRIS COLLINS

4 'WE COULD BE IN FOR A BLOW'

NOT LONG AFTER lunch on Christmas Eve, the meteorologists were certain that *Tracy* was destined to impact upon the Darwin coastline and had a very good chance of striking the city itself. In global terms, *Tracy* wasn't a big tropical cyclone. Geoff Crane could recall storms that evolved off the coast of Queensland that filled the entire Coral Sea. The gale-force winds had extended out to 600 kilometres, with winds above 40 knots through the Whitsunday Islands. Very big tropical cyclones have a lot more total energy, but the maximum winds are a function of how steep the pressure gradient is near the centre. It is possible to have a huge tropical cyclone that covers a very large area and which might have very low central pressures, but if the pressure falls over a long distance—across the diameter of the revolving storm—then the maximum winds might not be high. Wind strength in a tropical cyclone is a function of how steep the pressure gradient is leading into the centre of the storm.

By early afternoon on Christmas Eve, Nick Paspaley was still steaming towards Darwin with his wide-eyed crew. The *Paspaley Pearl* now had better seas—although still very rough in the relatively shallow waters—as she passed through the Vernon Islands south of the Melville Island landmass that offered protection from the winds. Nick's crew were recovering from their night surfing. They hoped to arrive in Darwin, which was now experiencing very leaden skies and persistent drizzling rain, by late afternoon.

Rick Conlon was still working at the hospital and telling everyone in his crew to head home, as 'there was nothing more they could do to prepare for the storm'. He thought the weather was 'like a fog with a runny nose'.

Down at the port, Colin Wood, the assistant harbour master, had rounded up all the trawler companies with boats in port and warned them to secure their craft. Those trawlers with crews left the inner harbour and most sheltered at Frances Bay, a sizeable mangrove-vegetated bay at the entrance to East Arm. East Arm extends for 20 kilometres and has several large creeks as tributaries. Frances Bay is also in the lee of Darwin Peninsula and looks up about 50 metres to the suburb of Stuart Park and the light industrial area of Winnellie. The immovable vessels in the harbour were tied up to the inside of the Stokes Hill Wharf. The steel-hulled harbour tugs *Goyder* and *Corowa* were kept alongside the wharf ready to assist in moving craft until the last possible moment. The wharf precinct was a hive of activity, with all cargo on the wharves being lashed down as best as possible. Carl Allridge recalled that the wind was up, but he had seen worse in a dry season south-easterly. The sky was cloudy and deeply overcast, and it was rough out on the water.

The storm closing in on the town didn't deter the meteorologists from having a Christmas work party in the amenities room at lunchtime. The operations people, who were now maintaining almost constant contact with all of their reporting agencies, were kept busy up on the eighth floor and couldn't attend. The party lasted until mid-afternoon and then everyone except the rostered staff went home.

Many other Christmas Eve work parties around town also finished earlier than usual, as the heavy rain persisted and people began to feel the looming presence of the cyclone. Cherry Perron, whose husband Marshall had just been elected to the newly formed Northern Territory Legislative Assembly, was working in the Qantas office in town. Like many other businesses, it had shut down for the day, except for those staff working out at the airport. The staff were having Christmas drinks. Because they were vitally interested in how the weather affected flight operations, everyone was acutely aware of the tropical cyclone. Cherry recalled someone saying, 'We could be in for a blow.'

Eddie Quong left his workplace and returned to his house in Shepherd Street in the city. While Greta continued with her cooking preparations, Eddie packed a radio and candles, but no clothes or matches. He decided that his two sons, aged seventeen and fifteen, would be best sheltered in a downstairs flat beside the house; he and his wife and their two daughters would stay upstairs. Eddie and the boys spent the remainder of the afternoon cleaning up the backyard and securing everything they could.

Over at the airport, Bill Gough had been busy securing everything at Con Air's maintenance hangar. The company had decided to fly its aircraft south to Katherine, some 275 kilometres south of Darwin, which would be well out of reach of any tropical cyclone. The crews of the aircraft would be flown back in one aircraft as soon as they had parked and checked in their planes. As it turned out, when the DC-3 aircraft bringing the flight crews back from Katherine attempted to land at about 5.00 p.m., it simply wasn't possible, owing to the strong winds and driving rain. Con Air still had one multi-engine, sixteen-passenger aircraft in Darwin, but it wasn't serviceable for flight and was secured in their service hangar. Out on the tarmac, the owners of dozens of light aircraft that were based in Darwin were securing their aircraft to tie-down points in the light aircraft parking area.

Tricia and Graeme Clarke lived out in the northern suburb of Wagaman, which was directly north of the airport. Graeme worked as a traffic officer for Trans Australia Airlines (TAA) in town, as well as at the airport when flights were due in and out. A former New Zealander, he had married local girl Tricia and they had two young children aged nineteen months and two months. Their house was formerly the local radio station 8DN staff house, which had been relocated to Wagaman. Tricia's family owned the house, and the Clarkes were living in it until their new house, which was being built in Vanderlin Drive, just 500 metres up the road, was completed. Their current accommodation was an old, typically high-set, two-bedroom cottage on steel piers with a long hallway down one side. It had a combined lounge/dining room, a bathroom and separate toilet, and, as was typical of high-set homes, an open laundry underneath.

Graeme left work at lunchtime and drove up Bagot Road in conditions he described as 'gusting and rainy'—and, as it turned out, 'the calm before the storm'. There wasn't a lot of traffic on the road at 2.00 p.m. and he suspected that most people had left their work parties to get home and prepare for the storm. Working for TAA had given Graeme a very good idea of what was heading for town. He set about securing the yard of the house they were living in, and then went up the road to check on their house that was under construction. The workmen were long gone. He tidied up the yard where bits of timber and other cast-off material was lying around. He was thorough and spent some time securing the place before returning home where Tricia was waiting. 'It was obvious that we were in for a blow,' she said. 'Radio warnings were very specific, and we made fairly good emergency preparations.' She recalled that people were advised not to shelter in their car or underneath their house. Before they sat down for dinner that evening, Graeme taped all of the windows in the old house.

By 3.00 p.m., very heavy rain had started to fall in Darwin and the wind had strengthened considerably. Tropical Cyclone *Tracy* was now only 80 kilometres from Darwin and heading directly for town. It was evident that it would be more than a normal wet season blow. The band of winds wouldn't be felt until *Tracy* was 35 to 40 kilometres from the coast. There was a band of gale-force winds about 25 to 35 kilometres from the eye of the tropical cyclone. Winds that could cause damage to buildings and trees could be expected around 25 kilometres out from the eye. Gusts within those winds that were likely to exceed 160 kilometres per hour were closer in, about 15 kilometres from the centre of the approaching storm.

Chris Kingston-Lee went to Christmas drinks at a car sale colleague's workplace in the CBD. 'The weather got ear popping,' he said. 'The sky was a funny colour, like bushfires had been burn-ing.' Ted, a local car detailer who had lived in Darwin all his life and worked for many of the car yards around town, predicted that *Tracy* wouldn't come, and everyone had another beer.

At Larrakeyah Army Barracks and the RAAF base, the street wardens were circulating and advising the married quarter tenants of what to do in the event of a tropical cyclone. At Larrakeyah, Diane Kerntke heard the warnings and, like the organised person she is, put a load of washing on in the laundry. It would have to dry later under the house, she thought, as she looked out to the west where the sky was heavily overcast and grey. Her husband Peter had been advised that he would now be working through Christmas Eve manning the operations centre. He had been called away from his all ranks party to set up the communications in the operations (ops) room. Peter had warned Diane that a tropical cyclone was headed for Darwin and that if it closed on the town she and their two children would be sheltered in the Officers' Mess building, 150 metres from their high-set house. That building, like the Sergeants' Mess and the Other Ranks' Mess, was a large two-storey brick and concrete building that had withstood being bombed by the Japanese during the Second World War. The Messes were the designated tropical cyclone shelters for those families that wanted to use them.

Up at the ops room, Peter Kerntke established communications by VHF radio with the RAAF and the disaster network was set up. On duty in the communications centre were two other soldiers and a staff officer from Headquarters of 7 Military District, Barry Darr. While Peter was setting up the ops room, he could sense that the situation was becoming serious, but he still wasn't 'all that concerned'. At home, Diane busied herself putting their paintings under the cushions of the lounge suite, planted the Christmas tree in the toilet bowl to keep it upright, and when Peter returned home for dinner they packed the Christmas presents into a suitcase.

John Ryan left work in the CBD and went down to a Christmas Eve party at the Sailing Club in Fannie Bay, near the Darwin gaol, where he had a few beers. Looking out over the ocean, he saw the large swell heaving under a black sky, and out in the boat parking area he could see people tying down their boats. At around 4.00 p.m. he decided to go home to Alawa. 'The signs just didn't look good,' he said. When he heard the warnings over the car radio, he decided he'd made the right decision.

Not far from where John Ryan was making his decision to cut short the bonhomie with his fellow Sailing Club members, Laurie Gwynne had been visiting the Fannie Bay gaol. Laurie worked for the Forestry, Fisheries and Wildlife Department and was a senior forestry officer. The Fannie Bay prisoners did work on the plantations for Laurie out at the Berrimah research station. He was at the gaol organising some details for work in the coming months. From the gaol he could see that, looking to the west, the sea was dark. 'I could hear the sea sort of moaning,' he said. Rain squalls were visible across the width of the panorama. Laurie was also an ex-meteorological officer and was very conscious of tropical cyclones. He had said to his wife Shirley many times that he thought a Darwin tropical cyclone would have to come from the north-west, and now it looked to be a fact.

Out in the suburbs, a number of people, including Judith Watson, were preparing for Christmas Eve parties and hoping that *Tracy* would be a storm like *Selma*. Judith's husband Gary was on his way home from a work party at Port Darwin Motors where he worked as a car salesman. He had listened to the same people that Chris Kingston-Lee had talked to at the party and he thought the tropical cyclone would veer away like they always do. Judith and Gary had planned a party in the backyard of their Wagaman house and were expecting over thirty people to attend. Judith had her weather eye out because she had some party hire people coming over with all the equipment for the party. They were due at 3.00 p.m., but when Gary said that he had spoken to Ted Brophy, the head of the Civil Defence Organisation—who, in turn, had been briefed by the mets from the Weather Bureau—things looked grim for a party to proceed. Ted Brophy was so convinced it would be bad, he had had his private plane moved out of town.

It was overcast, very grey and muggy. The Watsons had an above-ground pool in the backyard and the kids were enjoying a swim in the sticky weather. Tony, a friend, was mowing the grass in preparation for the marquee and chairs, but he advised Judith to call off the party. Gary arrived home just as heavy rain started and the decision was made to give the party a miss. They then got busy tidying up the yard, bringing in pot plants out of the open and putting the cover on the pool. They telephoned as many friends as they could contact to advise them that the party was off, and cancelled the booze delivery for the celebrations.

The next tropical cyclone warning at 4.00 p.m. should have convinced most people in Darwin that they were in for a blow.

PRIORITY. CYCLONE WARNING NO. 17 ISSUED BY THE DARWIN
TROPICAL CYCLONE WARNING CENTRE AT 4.00 P.M. CST,
24 DECEMBER 1974.
AT 3.00 P.M. SEVERE TROPICAL CYCLONE TRACY WAS CENTRED
80 KILOMETRES WEST-NORTHWEST OF DARWIN AND MOVING SOUTH-
EAST AT 7 KILOMETRES PER HOUR.
THE CENTRE IS EXPECTED TO BE NEAR GROSE ISLAND ABOUT
MIDNIGHT.
VERY DESTRUCTIVE WINDS OF 120 KILOMETRES PER HOUR WITH
GUSTS TO 150 KILOMETRES PER HOUR ARE EXPECTED BETWEEN
DARWIN AND THE PERRON ISLANDS TONIGHT AND TOMORROW
MORNING.

Kevin Jurek was foreman to a gang of men who were responsible for diesel power-generation supplies in Darwin. A work party at Winnellie had started after they had all knocked off for the day at around 4.00 p.m. The electrical gang were gathered in the workshop, since it was too wet outside. At 7.00 p.m. they conceded that it was time to head off home. Not far up the road from Kevin Jurek's group, young Nat Guinane's father was also attending a work party with about forty people from the Administrative Works Section 7, known as AW 7, who looked after road maintenance and other public works. The party was held in a large storage hangar. The men's families were present, and there was plenty to eat and drink. The party lasted several hours until 'the boss ordered us out' and told everyone to go home.

Many people returning home in the late afternoon did some final shopping before the Christmas break. Rowan Charrington and his wife Marcia went down to the shopping centre at Casuarina, not far from Wanguri where they lived. Rowan recalled that it was 'overcast, the sky was leaden with heavy clouds'. He felt 'suspicious of the sky, and had a weird—almost foreboding—feeling'. The atmosphere was 'unsettling', he said. 'There was electricity in the air.'

By 5.00 p.m. it was starting to get dark because of the low cloud. The sun normally set at around 6.00 p.m. at this time of year, but the overcast conditions, low cloud and heavy rain brought an early evening. Tropical Cyclone *Tracy* was now only 80 kilometres offshore and heavy winds were buffeting the coastline. Although not yet gale-force, they were very strong indeed. Trees were bending, and loose items such as rubbish bins were being tipped over.

Julie Tammo, in Larrakeyah, was worried because her four-year-old son, Rohan, was quite ill. She decided to call on her local GP, Dr Ella Stack (later to become mayor of Darwin), who lived just around the corner from the Tammos. Dr Stack advised Julie to get Rohan to the hospital as soon as possible. Julie's firefighter husband Heikki was off shift, so he looked after their other son while Julie went with Rohan to Darwin Hospital. Waiting at the children's ward at the rear of the main hospital complex, Julie could see the ocean from the cliff edge. It looked like a high tide, but she thought it was unusual because 'the sea was boiling and looked funny'.

At about this time, Nick Paspaley was entering Darwin Harbour. 'By the time we got to Darwin Harbour,' he said, 'the real angry seas were over, but it was still rough and choppy in the harbour.' Nick noticed that there were some boats on the inside of the main wharf, but nothing on the outside. All of the Navy patrol boats were out on their cyclone moorings. He could see that the boats were prepared for a storm. Nick decided that it was too rough to risk a berth alongside the wharf. 'I nosed my bow on to the end of the wharf and managed to get all the passengers off the bow and just dropped them there on to the wharf.' The harbour master had been advising people to go up into the creeks as far as possible and, if necessary, ground their craft as high on the tide as they could. That type of action is, of course, directly contrary to what most masters or captains would want to do to a vessel, but it was good advice.

Nick sailed out of the wharf area and headed east around the end of Darwin Peninsula and then north up into Frances Bay, passing the power station and the suburb of Stuart Park up on the hills above. 'I went up to Sadgroves Creek, which is a mangrove creek, in Frances Bay. I got up to the top of the creek as far as I could go with that boat and played out both anchors and chains. Then I got in my dinghy and went ashore. It was about 8 o'clock.' The tide was quite high, Nick said. Sadgroves Creek is only about 200 metres wide and so he anchored the *Paspaley Pearl* in the middle of the creek. After rowing ashore, he put the dinghy on the back of his ute, drove around to the Paspaley sheds, dropped it off in the yard that was near the waterfront and then made it home.

When Nick rushed into his house, he found that his wife 'just wasn't having the baby. It was as simple as that.' The last report Nick had heard had indicated that the tropical cyclone was heading west, so he wasn't aware that it was now headed directly for Darwin. 'It was pretty windy; it was probably gusting around 30 to 40 knots. So I said to my wife, "Let's go to my mum's place and see if they're all right." And we went over to Mum's place at Myilly Point.'

Tracy was now less than 70 kilometres from town. Somewhat amazingly, people were still arriving at Darwin airport for Qantas flights in the late afternoon and early evening. At the Casuarina shopping centre, forty-seven-year-old Jenny Lonergan was working in the large Coles store. She had been outside a few times during the day where, she said, she could 'hear the wind talking'. It got 'noisier with the gusts'. Jenny and her husband John, a foreman at the Darwin power station, lived at Humpty Doo, some 45 kilometres out of town, and she wanted to get home and square things away. She had been listening all day to customers in the store talking about the cyclone, so she knew the situation wasn't good. Her boss had wanted to keep the store open as late as possible for the trade, and so it wasn't until 5.30 p.m. that the staff were able to leave. John had picked up a new trail bike at lunchtime for Christmas but was unable to ride it, having over-indulged at his work party that afternoon. Jenny had to drive them both home.

While Jenny Lonergan was struggling to get her 'legless' husband into her car and get him home, Police Sergeant John Woodcock arrived at his home in Croker Street, Nakara, somewhat less inebriated. (He was a policeman, after all.) Nakara was the limit of the northern suburbs at that time and was next-door to Alawa and across the main arterial road, Trower Road, from Wanguri and Wagaman. John had attended the police Christmas drinks. It was a huge affair with about one hundred and fifty coppers and colleagues from different parts of the community in attendance. He had left while the function was still going, as he wanted to fix things up around the house. John and his wife Janice were buying their ex-government high-set house but hadn't yet signed the papers to finalise the deal. They were long-time Darwinians. John's elderly parents were up in Darwin on a caravaning holiday from the gem fields. They had parked their 8-metre van in the backyard next to the Woodcocks' own caravan. John was rostered off for the next four days and was looking forward to spending some time with his parents over Christmas. As he drove home down Trower Road, he noticed a large number of rubbish bins being blown around. Rain squalls made observation difficult, and driving took some concentration in the very wet, windy conditions. John was expecting high seas, some minor flooding and probably quite a few trees down as a result of the storm—like they had experienced with Selma.

As the sun was setting, the Charringtons across the way at Wanguri were in a flap. They had done the right thing and put out their absent neighbours' rubbish and handed a black plastic bag to the garbage men, only to find shortly after that in fact it was a bag full of their good clothes. The rubbish truck had disappeared down the road doing a last run before the holiday break and they were now contemplating spending Christmas Day at the dump looking for a bag of clothing.

At 6.00 p.m. it was dark enough to need the car headlights on to drive. *Tracy* was only 50 kilometres from Darwin. It was very windy and the rain was quite heavy, blowing in gusts that lasted for up to a minute. It was also becoming unseasonally cool. The wind noise was increasing.

In the suburb of Nightcliff, which overlooks the ocean, fifteen-year-old Lesley Mance's policeman brother had just left for night shift. Before he left for work he had briefed his wife, Rita, his mother and young Lesley on what to do if the tropical cyclone hit. He had also battened down his half-cabin boat on its trailer under the house and secured everything in the yard. They were ready. They had the radio on and at 7.00 p.m. heard the following warning:

```
TOP PRIORITY. FLASH CYCLONE WARNING NO. 18 ISSUED BY
THE DARWIN TROPICAL CYCLONE WARNING CENTRE AT 7.00 P.M.
CST, 24 DECEMBER 1974.
AT 6.00 P.M. SEVERE TROPICAL CYCLONE TRACY WAS CENTRED
60 KILOMETRES WEST-NORTHWEST OF DARWIN AND MOVING
EASTWARD CLOSER TO DARWIN AT ABOUT 6 KILOMETRES PER
HOUR.
THE CENTRE IS EXPECTED TO BE IN THE SHOAL BAY AREA AT
6.00 A.M. TOMORROW.
VERY DESTRUCTIVE WINDS OF 120 KILOMETRES PER HOUR WITH
GUSTS TO 150 KILOMETRES PER HOUR ARE EXPECTED BETWEEN
CHARLES POINT AND THE VERNONS TONIGHT AND TOMORROW
MORNING.
```

Shoal Bay is the body of water that sits between Shoal Bay Peninsula and Gunn Point. It is immediately north of where Lesley Mance and her mother, sister and baby nephew were living. No one said much, and Lesley was worried. The wind was beating hard against the high-set house, and rain was splattering hard against the glass and metal louvres in the lounge room where they were having a cup of tea after dinner.

Around the corner, Chris Kingston-Lee was going to a party held by a new couple in town in their flat. Chris arrived about 7.00 p.m. 'It was getting quite windy. I could see the sea and it was all white looking from the cliffs at Nightcliff.' His hosts were recently arrived from Adelaide and were talking about the storm. Cynthia, the hostess, commented that the storms in Spencer Gulf in South Australia were worse.

Two streets away, in Bougainvillea Street in Nightcliff, Nat Guinane and her family had just arrived home in the family's Valiant sedan from the AW 7 work party. They lived in a flat that was part of a set of four units. The flats were built of concrete blocks with a suspended concrete slab floor. The windows were louvres. The flat comprised a lounge/dining room at the front, two bedrooms, and a kitchen at the rear. They had also heard the 7.00 p.m. warning, and Nat's father started gathering mattresses to provide protection in the hallway. Nat became a little scared when she heard her father say, 'This is going to happen.'

At Fannie Bay, Milton Drew, who had only recently moved to Darwin, was having dinner with one of his company representatives, Ray Johnson. Milton was the manager of CIG (Commonwealth Industrial Gases), a privately owned company that supplied the bulk of gases to industry and commerce in Darwin. They had a large plant in the industrial area of Winnellie. Milton's family was still in Brisbane and he was living in an almost empty house save for a bed, chairs, table and, of course, a beer fridge. His furniture was in storage awaiting the arrival of his wife because, he explained, 'I wasn't going to put it all down in one place and only have to move it again when she arrived.' Milton wasn't too concerned about the storm, primarily because Ray Johnson wasn't. But when the wind became very strong at around 8.00 p.m., they decided to prepare for the night and Milton drove back to his house in Casuarina, about 6 kilometres away. Milton recalled that as he was driving up Bagot Road, his car was 'buffeted by the strong winds as I was driving along the road. It took some effort, and there weren't many cars on the road at that time.'

Kevin Jurek was also on the road at this time, returning from his work party and heading for Jingili, a northern suburb close to the northern perimeter of the airport. It was raining very heavily, which made it difficult to drive. He and his wife Karen were supposed to be going to a friend's twenty-first birthday party, but a combination of the storm and Kevin's sobriety annoyed Karen and they stayed home. Kevin fell asleep in a lounge chair and Karen tidied up around the house.

Jenny Lonergan had negotiated the road home to Humpty Doo in her four-wheel-drive, canvas-topped, Suzuki 'Mustard Monster'. Their home in Cypress Road was a low-set ranch-style house and on acreage. Jenny put her husband to bed without any tea, even though she had made him his favourite meal of garlic meatballs. Instead, Goldie, their pet Labrador dog, got to scoff the lot—a treat that would later backfire on them.

Meteorologist Geoff Crane had decided when he came off shift that afternoon that he would take his wife Barbara and their children back into town and shelter them in the MLC Building with the families of some of the other mets. He came home for tea at around 7.00 p.m. When they left soon after, it was quite windy and raining heavily. Barbara had been busy. She had taken down all the pictures from the walls and put them in cupboards. She had also placed some of the kitchen appliances in cupboards under the sink. On the way back into town from Rapid Creek, Geoff stopped off at the radar shack at the airport and picked up the latest radar plots.

Carl Allridge, the harbour master, was at home but in constant contact with the Weather Bureau. In his lovely home at Fannie Bay on the corner of Parsons and Ross Smith Avenue, he was only 500 metres from the ocean. The house had an iron roof, was set up on concrete piers and had banks of louvres to keep it cool. An enormous 9 x 9-metre balcony took advantage of the sea breezes and view of the sea. Underneath the house was fully concreted, and this area adjoined a 10-metre in-ground swimming pool. Carl was always hosting guests, usually ship's masters, and it was an ideal house for entertaining. It wasn't until Carl had been in town for about seven or eight years that he learned his house was on the tourist route, so beautiful were the gardens and setting.

By 8.30 p.m. the winds were strengthening considerably. Assistant harbour master Colin Wood, who was on duty down at the harbour, rang Carl and suggested that they place additional mooring lines on the trawler *Jenny Wright* because it appeared to be crew-less. Carl agreed, and Colin said that he would get the traffic officer at the port to assist him. Colin then decided to stay in his office down at the wharf because the wind had got up. He told Carl that the jetty was shaking. Later that night he couldn't have gone home if he had wanted to, as his car was blown over on its side.

At the RAAF base, Josephine Foreman was in her married quarters. The RAAF truck had been around several times after dinner warning people of the precautions they should be taking. Josephine's husband, who had been drinking all day, was 'smashed' and had gone to bed at 8.30 p.m. Josephine described the atmosphere as being like 'the earth was holding its breath before the first wind'. There had been a hot, heavy feeling in the air all afternoon, with a strange drizzle. Now it was raining incessantly, pelting down on the corrugated iron roof, which made it hard to hear the radio. She wanted to listen to Evening Mass on the radio and was hoping the rain would ease before then.

Tony and Judy Pickering lived in Bremer Street, Ludmilla, which adjoined the RAAF base, but they were destined not to spend the night in their own home. After tea, they gathered up their eighteen-month-old son, Ben, and attended a party around the corner in Wells Street, just a couple of minutes' drive away. Tony told their friends at the party what Geoff Crane had told him earlier in the day about the cyclone, but it did little to deter the boisterous party mood.

The tropical cyclone warnings on the afternoon of Christmas Eve had been issued every three hours until 7.00 p.m. These warnings had come from satellite reports. But from 9.00 p.m. until midnight—referred to as the 'critical stage' by the mets—the radar provided auxiliary warnings every half-hour.

```
TOP PRIORITY. CYCLONE WARNING ISSUED BY THE DARWIN
TROPICAL CYCLONE WARNING CENTRE AT 9.00 P.M. CST,
24 DECEMBER 1974.
TROPICAL CYCLONE TRACY WAS LOCATED BY RADAR
50 KILOMETRES WEST-NORTHWEST OF DARWIN A FEW MINUTES
AGO AND MOVING EAST-SOUTHEAST AT 6 KILOMETRES PER HOUR.
```

The Berrimah Hotel had a disco night going full steam in the large lounge area, where eighteen-year-old Peter McIver and his mates were having a great time. But the pub closed early, and so they headed off back into town up the Stuart Highway. Peter's mate was driving, and he recalled that when they stopped for fuel at the Shell servo it was 'blowing a gale'. He admitted that they were 'a bit pissed', but he said the drive back into town smartened them up. Branches, palm fronds and leaves were blowing across the road, and loose rubbish was whipping past the car.

At 9.00 p.m., Captain Peter Kerntke was back in the army ops room at Larrakeyah Barracks. The ops room was a solid, double-brick, low-set structure with an iron roof. It had no windows. Peter's wife Diane was at home with their kids and the neighbours' dog. She spent some time doing her nails for the Christmas party they were hosting the next day.

Peter Kerntke and Barry Darr had been monitoring their street warden system, which comprised soldiers touring the barracks and married quarter patch with VHF-25 radio sets. With this internal network, they were capable of coordinating activities within the base even if the telephones went out. They still had communications with the RAAF and Police Incident Room in town. Peter was plotting the tropical cyclone on the operations map and knew it had passed the top of Cox Peninsula, the peninsula on the western side of Port Darwin. Like many others, he was hoping it would hit that landmass and weaken and then change direction—but it didn't.

Just around the corner at Darwin Hospital, Julie Tammo had

stayed at the children's ward with her son Rohan until 9.00 p.m. When he had fallen asleep, she had returned home to her house and waiting husband in very blustery conditions. The wind was now beginning to make a lot of noise. Julie was extremely nervous and occupied herself by gathering up the Christmas presents and putting them in the spare bedroom. As set out in the tropical cyclone instruction handbook, she then put water in the bath and assembled some emergency provisions. They didn't tape the windows because they were mostly louvres, and they wanted to be able to release the wind and air pressure once the storm changed direction. Heikki was on shift the next day, so they went to bed at around 10.00 p.m. They were looking after Julie's mother's dog while her parents were on holiday in Adelaide, and the animal was in a small shed under the house. Because of the hot nights in the wet season, most people in Darwin slept naked. But on this night, because it was a little cooler, Julie wore 'a very thin nightie, and my hubby wore undies'.

The inner suburb of Kahlan is nestled down behind some large hills just to the north of the CBD and fronts down close to the water at the back of Larrakeyah Barracks and Myilly Point. Today, a tropical cyclone-proof marina with a storm lock is located in that same area; it's called Cullen Bay and is a prestigious part of Darwin. Grant Tambling (another recently elected member of the Northern Territory Legislative Assembly and today a federal Senator) and his wife were hosting a Christmas Eve dinner party and had friends around until after 10.00 p.m. They heard the tropical cyclone warning on the radio about that time and realised that things were getting serious. They were protected to a considerable degree down in Kahlan and weren't aware just how strong the wind was until it was time for people to go home. Grant's sister had to drive back to Jingili, and the fifteen minute drive took her ninety minutes in the driving wind and rain.

At Fannie Bay, Elspeth Harvey had been busy preparing food for a traditional Christmas lunch and, as she said, 'driving my husband Alexander and daughter Gail mad singing Christmas carols'. Their son lived in a flat just around the corner. The Harveys went to bed early, with Elspeth feeling especially pleased with the new 1.5-metre Christmas tree she had bought and decorated. It glistened beautifully with all its new lights and baubles.

Out at Judith and Gary Watson's cancelled party at Wagaman, the wind was now very strong. A friend called Stephen (whom they had been unable to contact to say the party was cancelled) had turned up, and they had been having a drink or three with him. When Stephen called a taxi and tried to walk out to the waiting vehicle, the wind pushed him off his feet and he had to crawl the rest of the way.

Chris Collins at Nightcliff realised also at about this time that his caroling friends weren't going to call. He and wife Tricia had spent the evening 'listening to music, reading and waiting for the singers'. They reckoned that they probably wouldn't turn up, because they wouldn't be heard above 'the din of the shrieking wind and rain that was now lashing down'. Chris had put aside some liquor for the carolers' Christmas cake, but had used it himself. He had taken several precautions around the house, including placing large tarpaulins over his cherished stereo equipment and the double bed. He had taped all of the windows, and had even filled his dinghy under the house full of water to stop it moving around. Chris and Tricia were planning on going to Tricia's parents' place for lunch on Christmas Day around the corner in Pandanus Street.

Chris watched out the window while waiting for his friends and saw that vehicles were having trouble negotiating the roadways in the strong winds. It was now very dark, with heavy torrential rain coming down at about a thirty-degree angle to the ground. A brave motorist in a Holden FB station wagon was trying to drive up the street. The car would get level with Chris's house and then literally be blown back down the street. The driver's motoring endeavour was illuminated by flashes of lightning, which had started to increase in intensity, and by the streetlights, which were still working. After three attempts to get up the street the driver gave up, and Chris didn't see him or his car again.

```
TOP PRIORITY. CYCLONE WARNING ISSUED BY THE DARWIN
TROPICAL CYCLONE WARNING CENTRE AT 9.30 P.M. CST,
24 DECEMBER 1974.
TROPICAL CYCLONE TRACY WAS LOCATED BY RADAR
46 KILOMETRES WEST-NORTHWEST OF DARWIN A FEW MINUTES
AGO AND MOVING EAST-SOUTHEAST AT 6 KILOMETRES PER HOUR.
```

Nick Paspaley was at his parents' place at Myilly Point, an extremely exposed finger of land that juts out into Cullen Bay. He had taken blankets, a torch and emergency supplies with him. Despite the raging storm, he was feeling 'a bit relaxed, because I had got off the ship. My heart and the adrenaline had been running all day, and now I felt as safe as a house with my feet on the ground. I was home and I felt that the boat was safe.' Nick wasn't really worried about the tropical cyclone. His father was a tough one and wasn't the least bit concerned. Nick recalled, 'He turned around at one point and said, "I don't know what all the fuss is about." He went off to bed about 9 o'clock.'

At about the time his father retired for the night, Nick got a phone call from Darwin Radio, the ships' radio. They passed on a message that someone on his yacht, who was also anchored in Sadgroves Creek, reported that the *Paspaley Pearl* had dragged anchor and they felt that it was going to drag down and damage them. Nick was asked to go and move his boat. In Nick's words, 'When I heard that, I thought, "Fuck! That doesn't sound too good." So, I told my old man that I was going to go down to move the boat and make sure it doesn't drag and then I'd be back.' Nick grabbed his brother-in-law who was innocently standing by and they hopped in Nick's ute. But just as they were about to leave, the power in the house went off. Some trees had come down and taken down the power lines.

As Nick was driving out of the front yard, a live cable tangled around the car side-door mirror. It was sparking on the mirror and Nick thought, 'This doesn't look too good.' He couldn't get out and risk electrocution, so he drove off and ripped the mirror off the car. All the way down to the waterfront, the two intrepid boat movers encountered fallen trees across the road. Between 8.30 and 9.30 p.m., the wind had picked up considerably and was starting to do a reasonable amount of damage. The driving rain and rows of fallen trees were making it difficult for Nick to get to the water-front. The two men did a U-turn, headed back towards town, and then tried to go down behind the fuel farm, where there is a track that goes down through the mangroves towards Sadgroves Creek. Since he couldn't get to his dinghy, which was down at the water-front, Nick had decided to swim out to the boat. 'It was an easy

100-metre swim from the side of the creek.' But Nick admitted later, 'It was a pretty ambitious idea at the time.' When they reached the bridge that would take them down into the back of the other side of the waterfront, Nick was having problems.

> The car started to lose control and the wheels were lifting off the ground. So we backed back and took the back track down past the oil farms. I thought we'd make the waterfront, because once you got down amongst the mangroves it was a bit calmer and there was a cliff there which kept us out of the wind. We went across over the railway line and it was pretty muddy on the dirt track. We still thought we were going to be able to swim to the boat. But the high-tension power lines from the powerhouse were loose, and we saw this great big fire and spark off the end of the wire. There was a blue spark as big as a dining table and it exploded every time it hit something.

Nick and his brother-in-law thought they would be killed if they tried to get past the power lines. Nick didn't know what voltage the wires were—he reckoned maybe 100,000 volts—but they were the main lines. He figured that it would have just melted his car—and him. So, they turned around and decided to go home. It was worse returning to Myilly Point. All the trees were now down and they thought they were going to have to walk. Eventually, they made it home. But the battle was just about to begin.

Above where Nick Paspaley had been watching the power lines arcing through the air, the Huddys in Stuart Park were making final preparations for the storm. Greg Huddy's cousin had been through Cyclone *Althea*, which had torn into Townsville, in north Queensland, three Christmases earlier. He had told him to secure the cupboards and built-in wardrobes, so Greg nailed them all shut. He then realised that they had a rubbish bin full of smelly prawns, and he tried to drive up the road to dump the prawns as there would be no rubbish collections for another week. He realised the futility of his efforts when he came up on to the highway, and quickly returned home. From their rectangular, high-set three-bedroom house, Greg and Barbara could see that Frances Bay, down below, was being whipped into white foam. They made up an emergency kit. Like most people at Yuletide, they had a freezer full of food and

Greg put a big water cooler in the freezer to make a very big ice block—it lasted a week.

While the Huddys were being hit by strong winds, a friend at Nightcliff wasn't yet feeling the effects of the cyclone and invited the Huddys over for a drink. They declined. Police cars were now driving past the house, but they could only see the flashing lights because the rain was so heavy. On the coastal fringe it was really blowing a gale, and there were massive amounts of thunder and lightning. It was the Huddys' first tropical cyclone. They spent a couple of hours running around mopping up water from rain coming through the banks of louvres, but they gave up when they realised their efforts were in vain. Greg said, 'We went to bed about 10.00 p.m. Water was sloshing in the toilet bowl. The house was moving and it was quite eerie.'

In the same suburb and just around the corner, Marshall and Cherry Perron lived in a house in Gothenburg Crescent overlooking Frances Bay. They were listening to the ABC broadcast at 9.30 p.m. and heard the announcer say that *Tracy* was 46 kilometres north-west and travelling south-east. They immediately set to preparing for the worst. Marshall even tied the cupboard handles together so that the doors wouldn't swing open. Their house in Stuart Park wasn't a normal Darwin house. It was built into the side of the hill. It was still an elevated three-bedroom house, but it had lots of timber-framed sliding glass doors to take advantage of the views over Frances Bay. All these glass doors were carefully taped. There was an open laundry on a slab underneath the house. Marshall was a car enthusiast and a Thunderbird, an S-Type Jaguar and a Datsun 180B (Cherry's car) were parked under the house. Once they realised that Darwin was directly in the path of the storm, Marshall and Cherry changed into jeans and had slip-on shoes handy. Cherry moved their Christmas presents into a spare room that was close to the back of the house in towards the hill. She put water in the bath, but didn't think it was going to get really bad. They went to bed—'more for a lie down than for a sleep'—in their clothes at around 10.30 or 11.00 p.m. and took the radio with them. They couldn't sleep anyway, as the weather was now quite frightening: it was noisy, windy and rainy.

Frequent flashes of sheet lightning filled the sky, and the house was being buffeted.

Many residents of Darwin were now experiencing the buffeting of their houses. The high-set houses, in particular, now had serious winds pushing underneath the dwellings, and people like John and Sue Ryan in Alawa found the noise extremely disturbing. Their house in Watson Street was shaking and moving on its piers. John recalled, 'The house was groaning. Small debris was being blown around and banging into cars, fences and the side of the house and the downstairs laundry. It then got louder, especially when some Stramit roofing was ripped off a school building and banged against a power pole for hours before the wind really hit.'

The strong winds were now bringing down power lines, and by 10.00 p.m. many suburbs were experiencing power blackouts.

```
CYCLONE WARNING NO. 19 ISSUED BY THE DARWIN TROPICAL
CYCLONE WARNING CENTRE AT 10.00 P.M. CST, 24 DECEMBER
1974.
AT 9.30 P.M. SEVERE TROPICAL CYCLONE TRACY WAS CENTRED
41 KILOMETRES WEST-NORTHWEST OF DARWIN AND MOVING EAST-
SOUTHEAST AT 6 KILOMETRES PER HOUR.
THE CENTRE IS EXPECTED TO CROSS THE COAST IN THE
VICINITY OF DARWIN IN THE EARLY HOURS OF TOMORROW
MORNING.
VERY DESTRUCTIVE WINDS OF 120 KILOMETRES PER HOUR WITH
GUSTS TO 150 KILOMETRES PER HOUR ARE EXPECTED TO
CONTINUE BETWEEN CHARLES POINT AND THE VERNON ISLANDS
TONIGHT AND TOMORROW.
```

John Auld started his shift at the airport's Qantas terminal building shortly after 10.00 p.m. He was listening to the radio for the tropical cyclone warnings and decided to bring his Volkswagen car into the building. He had been asked to do a tarmac inspection for

MacRobertson Miller Airlines, whose pilot must have thought they had some chance of landing at Darwin. The winds were far too strong to attempt a landing, however, and the aircraft returned to Katherine. It was too dark to see winds on the airstrip. Six staff were on duty and a handful of passengers waited in the lounge, some wanting to know whether their flights would still be departing!

The airport terminal was on the other side of the airstrip from the RAAF married quarters about 1.5 kilometres away. Dianne Ferguson had just finished taking all the paintings off the walls and putting them in cupboards. At Nakara, the Woodcocks now had rain coming through the cracks in their closed louvres and flooding the floor and their new carpet. Janice Woodcock was pulling sheets and towels out of her linen press trying to soak up all the water that was flooding in. Under her breath she was cursing the 'bloody builders'. There was now a lot of water in the bedrooms, so John got everyone up and his parents out of their caravan downstairs; the cat and the dog needed no coaxing to come inside. Everyone congregated in the hallway while the wind and rain raged about the house.

The Woodcocks weren't alone in trying to mop up water. Across in the next suburb, Kevin Jurek had been kicked out of his lounge chair where he had crashed earlier, and he and Karen were also mopping up. Water was coming in through the louvres and under the eaves. The wind was now so strong the glass louvres were bending. Everything in the house was getting wet. Through the flashes of lightning, they could see their next-door neighbour taping his windows. They didn't get around to doing theirs as they were fully occupied trying to contain the water that was flooding into their low-set house.

```
TOP PRIORITY. CYCLONE WARNING ISSUED BY THE DARWIN
TROPICAL CYCLONE WARNING CENTRE AT 11.00 P.M. CST,
24 DECEMBER 1974.
TROPICAL CYCLONE TRACY WAS LOCATED BY RADAR AT 10.30 P.M.
37 KILOMETRES WEST-NORTHWEST OF DARWIN MOVING EAST-
SOUTHEAST AT 6 KILOMETRES PER HOUR.
```

Chris Kingston-Lee had been to a mate's place at Nightcliff for a party. When he arrived home at around 11.00 p.m., he said, 'Things were getting really windy. Dogs were yelping and barking and going nuts down the road. I started to feel uneasy, and the hair was starting to stand up on the back of my neck.' While driving back to Rapid Creek, he had had to dodge rubbish bins scattered all over the road and branches that had been ripped off trees. 'Lots of people were driving—probably going back home.' Chris and his wife Roslyn had planned a trip overseas, and their airline tickets and a wallet with $1500 in cash was sitting on the dresser in the lounge room. Just before midnight, Chris heard on the radio that *Tracy* was about an hour away. The wind tearing up over the cliffs above Doctors Gully was getting stronger.

Around 11.30 p.m., when the electricity wires started stripping off the power poles at Ludmilla, Judy and Tony Pickering were still at the party in Wells Street, just around the corner from their home in Bremer Street, Ludmilla. The radio was on in the lounge room and they became acutely aware that Darwin was going to be hit. They drove home in their Holden car around midnight or just before, packed a suitcase, and then drove in terrible wind and rain to friends in Clancy Street, Fannie Bay, five minutes away. Their friends' house had a fully bricked-in entertainment area under the house that Judy described as being 'like a bunker'. They felt it would offer the best protection.

```
TOP PRIORITY. CYCLONE WARNING ISSUED BY THE DARWIN
TROPICAL CYCLONE WARNING CENTRE AT MIDNIGHT CST,
24 DECEMBER 1974.
TROPICAL CYCLONE TRACY WAS LOCATED BY RADAR AT
11.30 P.M. 33 KILOMETRES WEST-NORTHWEST OF DARWIN
MOVING EAST-SOUTHEAST AT 6 KILOMETRES PER HOUR.
```

John Auld was monitoring the radio and the Midnight Mass being broadcast on the ABC network. In the background, he could hear the involuntary pealing of the church bells and cyclone noises roaring over the top of the broadcast. While John was on duty at the airport, his wife Helen, in Bald Circuit at Alawa, was now frightened out of her wits. Their house was particularly exposed, and just after midnight the glass louvres in their main bedroom blew over

Helen and the baby. The side of the house facing Casuarina blew in first, blowing in the windows in the bedroom on one side. Helen had shards of glass from the louvres stuck in her legs, 'but the baby was okay'.

At Howard Springs, 30 kilometres down The Track, the wind was strengthening. Rick Conlon and his family were watching the sky blaze almost continuously with sheet lightning. Rick and his wife stayed in their rumpus room and held a prayer session. After midnight they heard a huge bang—the front bedroom of their house had collapsed. A neighbour's roof had been ripped off and become airborne, and had taken out the end of their house. Rick moved everyone, including the dog, into the rumpus room. He gathered mattresses to wrap around them and to place over the kitchen table to provide some protection. And then they waited.

5 THE ANGRY SEA

LAURIE COFFEY HAD made his home in the Top End when the Second World War ended, working at a variety of jobs including ship repairs and salt farming for the uranium mines. In late 1974, Laurie was married with five children and lived in Houston Street, Larrakeyah. He worked as a marine engineer with a company called Barge Express, which was owned by John Grice and which specialised in moving stores and equipment to the remote island and coastal communities throughout the region. The barges had a very shallow draft of around 1.6 metres that allowed loading and unloading on the beaches—a procedure known as self-discharging. The sides of the barges were high to account for rough weather, especially during the dry season when blustery south-easterly winds could make life at sea a tad uncomfortable.

About ten days before Christmas, Laurie Coffey had done a supply trip to Amnipari in West Irian. He had agreed to do the trip on the condition that he would be back in Darwin to spend Christmas with his family. It would be his first Christmas at home in about eight years. The barge that Laurie was working on was called the *Allana Fay*. The crew had had some mechanical trouble both at Amnipari and on the return trip and reached Darwin a day late. They then had another contract to fulfil, which was to take a load of Christmas supplies down the coast to Kalumbaru Mission in Western Australia. Kalumbaru is between Wyndham and Derby on the Bonaparte Archipelago and about 500 kilometres by sea south from Darwin. John Grice had told 'Snowy' Roberts, the skipper of the *Alanna Fay*, that the cargo was packed and ready to go. He had promised the crew that they would be back in Darwin 'easily by

Christmas Eve—no problem at all'. As John's five-year-old son Philip was aboard for the trip, the crew were reassured that they would be back in Darwin well in time for Christmas. Laurie Coffey related the story of what happened next:

> So, we went to Kalumbaru, unloaded all the equipment, went to leave when the tide turned, and the ramp on the barge wouldn't lift. The hydraulic equipment had broken down. I couldn't repair the gear to get the ramp up. The people from the Mission had already taken all of their supplies and gone back inland. We finally got in touch with the Mission the next day. They got a front-end loader down, another front-end loader lifted the ramp, and we chained it up. We could only get out of Kalumbaru on a high tide, and by this time we had lost about three tides.

By that stage, things on the weather front weren't too bad. It was still early on Christmas Eve. They'd lost quite a bit of time, but the skipper reckoned he would have the boat pulling into the Barge Express Wharf at Frances Bay late that night—still in time for Santa to call on young Philip.

As the *Alanna Fay* steamed north back towards Darwin, however, the effects of Cyclone *Tracy* became very apparent. By late afternoon, the barge was just off the coast near Napier Broome Bay, about 330 kilometres south of Darwin. It was ploughing into huge seas, what Laurie Coffey described as 'a very big swell. It was that big a swell that it wasn't safe to go any further.' The shallow draft and blunt nose of the steel-hulled barge would have made the conditions very unpleasant, not to mention dangerous, for the crew. Skipper 'Snowy' Roberts decided to head for the beach. For most boats it would be a problem, but for a shallow-drafted barge that was *designed* to drive up on to a beach, it was a piece of cake. The hardest and most dangerous part was turning side-on to the swell and driving fast for the beach. Laurie Coffey continued:

> The weather was getting worse. We found out [over the radio] from a stand-by barge on the oil rig about the cyclone. The weather got that bad that the skipper just got us into a small bay in time and rammed the barge up in the mangroves—up on the beach. And we stayed there that night, and got away

the next day [Christmas Day] early in the morning. The weather was still fairly bad.

The crew had heard nothing about Cyclone *Tracy* since the radio conversation with the stand-by barge on the oil rig, probably the *Lady Cynthia*, despite monitoring the radio all night. Laurie said, 'We heard nothing at all.'

By late afternoon on Christmas Eve, the *Clipper Bird* was under way. Before dark, they had anchored up, using their own kedge, or anchor. Rob Perkins had been on a trawler for almost a year and could recall only one other time when they had used an anchor: when they were on the Barrier Reef and the crew went ashore to an island.

The crew set about preparing their evening meal, and the designated cook made a hearty stew. At about this time, the wind really picked up and very heavy rain started to pelt across the water. Many of the crew quickly lost their appetite—or their dinner—shortly afterwards. No one drank any beer—it stayed in the cartons. Everyone on board sensed that they would be in for a long night.

The trawler stayed on its anchor for several hours, but as the night wore on it became obvious that they would have trouble staying on station just by kedging, or dragging an anchor. Rob recalled, 'The skipper got a bit panicky after dark.' He ordered the crew to throw out the whole drum of cable on the anchor in an effort to stop the boat from drifting. By the time the cable was spent, there was 500 metres of it on the anchor, but the boat still wasn't holding on the spool.

By 10.00 p.m., things were getting pretty bad. Of the ten crew on board, some got very sick; the part-time deckhands on their university semester break were 'really crook'. Rob was called up to the wheelhouse and said he 'got scared when I saw that everything that guided the trawler had shorted out from the horizontal rain driving into the wheelhouse'. During the night, the skipper

ordered Rob to go aft and below decks to get the engineer, who
was tending the diesels. As he descended from the wheelhouse, the
wind tore at his wet weather gear and blinded him when he turned
into the gale. When he crossed the aft work deck, several 44-gallon
drums, other pieces of equipment and chains which had broken
loose from their tie-downs were hurtling backwards and forwards
across the wooden deck. They were lethal missiles which, if they
didn't kill him, could knock him overboard.

Rob tried to see through the pitch darkness to the hatch that
would take him below-decks. The rain was blinding; everything
was a blurry outline. The wind was now screaming. Rob waited
until the trawler pitched and he heard the drums and chains crash
against the lower gunwale, and then ran for his life across the
heaving deck. He dreaded slipping on some loose chain and then
being crushed when the gear rolled back on the next pitch. He
scuttled like a crazy crab across the deck, grabbed the hatch and
then almost fell down the ladder. He told the engineer that the
skipper wanted to see him, and then watched his face turn pale.

The two men struggled back through the gale to the wheelhouse.
Rob looked out at the huge, chopped-up swells they were riding.

 The water was just foam. There was about a metre of foam on the waves. The
spotlights were on, but I could only see to the next wave. All the running
lights were on. The skipper had the engines going flat out just to steam on
to the anchor. But at full pelt we would only manage about 10 knots.

Trawlers aren't made for speed, but for dragging enormous
weights through the water. The two big Caterpillar engines produced
a lot of torque. The skipper kept the *Clipper Bird* into the wind and
waves, trying to maintain a heading on to his anchor cable. They
were several kilometres from shore. Sometime after midnight the
trawler lost its radar; the scanner wouldn't turn owing to the gale-
force winds. New paint on the wheelhouse from the refit only two
weeks before was being stripped off the boat. Everyone on board
was wearing life vests. 'They weren't only handy in case the trawler
sank,' Rob explained. 'The padding also protected us from falling or
flying objects and when we were tossed around in the galley.'

The crew were a mixed bunch. Rob recalled that they all reacted differently to what was undoubtedly a terrifying experience. 'Some were cool. Some were just seasick and went to sleep, because they were still half-pissed from the pub! After 9.00 p.m. the sea got really rough, but we weren't pitching and tossing except when we were trying to manoeuvre.' Occasionally during this time, when at anchor, the *Clipper Bird* had to dodge other boats that loomed up out of the darkness or were suddenly lit up by sheet lightning. Once the radar went, the skipper had to rely on his compass. Rob recalled that 'it was almost useless trying to navigate in the dark'. When they had to manoeuvre to avoid another craft in the harbour, they were severely tossed about. Everyone on board felt miserable and scared. Rob continued:

 We bashed around the harbour for a few hours after midnight with only a compass, and then we lost the anchor. Sometime around 4.30 a.m. we started to bounce up on to rocks—we had no idea where we were. In fact, as it turned out, we were near the army barracks at Larrakeyah—near the new patrol boat base—just below the 30-metre tall cliffs. We couldn't see the cliffs because of the heavy rain. The skipper ordered the crew to launch a Beaufort life raft, but it was just too dangerous. The *Clipper Bird* was tipped over on a 45-degree angle. We just sat there hoping we wouldn't get blown around too much.

The vessel had taken water in one engine down through the exhaust and the boat had then lost most of its power to manoeuvre and make headway. But it was a moot point. They had been blown at least 6 to 8 kilometres around the harbour, and that was when the cyclone had passed over Darwin and the wind direction was from the south-east. Now the trawler was grounded in shallow water in a raging wind. Underneath her were mud, rocks and not much else. Rob recalled that 'the skipper got emotional afterwards—it was the first time he had lost a boat'. But he hadn't done too badly, compared to many other boats that had been in the harbour that night. As Rob Perkins said, 'We had a few dents, no holes, and water damage to an engine. More importantly—no lives were lost.' About an hour before dawn, the wind began to abate and the screaming died to a loud moan.

HMAS *Assail*, one of the RAN's four Attack Class patrol boats stationed in Darwin, was commanded by Lieutenant Chris Cleveland. Twenty-seven-year-old Chris had been a naval officer for seven years and had served in Viet Nam aboard the destroyer HMAS *Perth* in 1970–71.

At 10.00 a.m. on Christmas Eve, the *Assail* was alerted that a Japanese trawler, the *Kempira Maru*, was in trouble in the raging seas off Bathurst Island. The captain of the *Assail* and his crew were placed on alert as the stand-by rescue vessel in case a rescue needed to be launched. Lieutenant Chris Cleveland recalled watching the progress of Cyclone *Tracy* that day; he knew that he wouldn't be able to go to the aid of the trawler.

At around 2.00 p.m., the commander of the patrol boats, Captain Eric Johnston, ordered the four boats to go to their cyclone moorings. By 5.30 p.m., the *Assail* was secured on to her buoy. As *Tracy* smashed its way on to the Australian coastline the seas were whipped into a foaming fury. Operating the patrol boat in these extreme weather and sea conditions requires a great deal of seamanship and a highly trained crew. The Attack Class boats were very powerful, with a high power-to-weight ratio and had the considerable advantage of twin screws. The standard practice was to use the anchor cable (chain) to secure the craft to the buoy. Chris Cleveland described the conditions and his intentions:

> By about 1.00 a.m. we had started to drag and I had no alternative but to try and slip from the buoy. I had just finished briefing the foc'sle crew and rigging safety lines in order to cut the cable . . . then it parted anyway.

The patrol boat was heaving and straining on its mooring, but the pressure on the vessel and the securing lines was so great that at 2.00 a.m. the steel cable broke. The plan, then, was for the *Assail* to head out of the harbour and ride out the storm on the open sea. The patrol boats were extremely sea-worthy vessels and, being navy boats, were completely watertight. All openings had been closed as part of the preparations. The ship was conned from the wheelhouse by Chris Cleveland. His executive officer, Sub-Lieutenant Bruce Wilson, manned the radar station and fed readings to his skipper. The chief engineer, Chief Petty Officer Rowe, was on the engine controls and two lookouts on the flying bridge flanked the group. This form of manning is called Special Sea Dutymen. Down in the engine room the second engineer and chief stoker were nursing the motors; they were well prepared for operating in extreme conditions, when maintaining a flow of cooling water to the engines can cause enormous difficulties when the vessel is far from stable. The remainder of the nineteen crew was in the aft mess. Chris Cleveland's main priority was to make sure he didn't collide with one of the fifty or so other vessels in the general anchorage area. He recalled:

 We crash-started the engines when the cable broke and tried to leave. However, the initial wind direction laid the ship on her beam-ends twice, and I feared that further attempts would capsize her. The *Assail* was rolling at least 80 degrees, and the navigation sidelights rolled into the water several times. Visibility was now about 1 metre. The propellers were as much out of the water as they were in, and the echo sounder seemed to be pinging on the wave tops. The combined force of the wind and rain stripped the foremast of its top coat of paint . . . There was solid water half-way up the superstructure. One wave lifted the stern out of the water and another smashed one propeller.

The skipper decided that his best course of action would be to try to run what he called a 'racetrack pattern' between Mica Beach and Doctors Gully, directly into and then away from the howling winds. For the next two hours the *Assail* steamed up and down the harbour trying to avoid other craft that were being smashed and savaged by the tropical cyclone's fury. Chris couldn't determine the wave height, as there was no clear definition between the air and the sea. The lookouts had trouble seeing more than a couple of metres. By using his engines, Chris was able to run into the wind and manoeuvre. He was totally focused on reading the weather and what it was doing to the patrol boat, and his task was to pick a suitable time to turn at the end of each 'racetrack'—at the same time trying to ease the battering on his vessel and crew. He was well-prepared: he had de-tuned his radar and prepared what is known as a blind pilotage plan, which enabled him to steer safely in the harbour. Despite the ferocity of the wind and the waves, the patrol boat eventually made it out of the dangerous confines of the harbour and headed out to sea. They had to fight not only the elements, but also a 20-centimetre hurricane hawser that had wrapped around the port screw. Around 4.00 a.m. on Christmas morning, the wind veered rapidly to the north-west and Chris Cleveland and his crew were able to steam towards Bathurst Island.

Lieutenant Paul de Graaf, commander of the patrol boat HMAS *Attack*, recalled the wind whipping up the seas tremendously. In his time in the RAN, he had never seen seas like them. By nightfall, the sailors had estimated the wind velocity to be in excess of 170 knots, and 'mountainous seas were like crashing dumpers'. The *Attack* was in danger of being ripped from its buoy. Paul de Graaf explained:

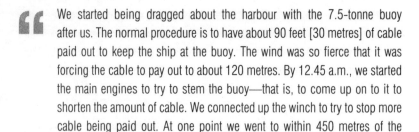

> We started being dragged about the harbour with the 7.5-tonne buoy after us. The normal procedure is to have about 90 feet [30 metres] of cable paid out to keep the ship at the buoy. The wind was so fierce that it was forcing the cable to pay out to about 120 metres. By 12.45 a.m., we started the main engines to try to stem the buoy—that is, to come up on to it to shorten the amount of cable. We connected up the winch to try to stop more cable being paid out. At one point we went to within 450 metres of the wharf. I managed to get us back to about 1000 metres.

The *Attack* was dragging gradually in a south-westerly direction as the leading edge of the tropical cyclone tore into the port. The biggest problem for de Graaf was avoiding collisions with the eighteen or so trawlers in his immediate vicinity in the inky blackness. The only ambient light was provided by the flashes of sheet lightning illuminating the raging sea. The patrol boat lost the use of its gyro compass very early in the night, and the crew were only able to steer by using a magnetic compass which, according to de Graaf, 'was totally inaccurate in the rough weather and practically useless'. Sometime during this frightening night at sea, *Tracy* passed over Darwin and the wind changed direction; it now blasted in with what seemed even greater fury from the south-east. The *Attack* was now being driven in a north-easterly direction— and closer to the shoreline. By about 4.00 a.m., the ship's radar stopped working and they were now totally blind. The patrol boat was still attached to its buoy, which was dragging through the muddy sea bottom, and somehow still avoiding prawn trawlers. Disaster then struck, as Paul de Graaf explained:

> We had had it when a prawn trawler got to within 7 metres of us. It was while trying to avoid her that the patrol boat ran aground. And just before we went aground, the mooring cable parted at the bows. We all stayed on board until the cyclone moved away. Unbelievably, we were all safe.

Their patrol boat—which normally drew 2.2 metres—was now high and dry in Doctors Gully on a muddy, debris-strewn beach some 4 to 5 kilometres from where it had attempted to stay at its mooring buoy. When it was eventually driven ashore after the mooring line snapped, it suffered three holes in the hull but was able to be refloated and return to the harbour. The prawn trawler that brought the *Attack* undone may well have been the *Clipper Bird*, with Rob Perkins on board, or the unmanned *Jenny Wright*, which sank around 60 metres away.

Fortune smiled on the patrol boat HMAS *Advance*, which was commanded by Lieutenant Peter Breeze. He had also been tracking the tropical cyclone on his ship's radar all afternoon and thought that *Tracy* looked 'like a bad one'. When ordered to tie up at his mooring buoy, he decided instead to anchor in the harbour, because he was 'scared that if the worst came to the worst, we might not be able to get ourselves away from the buoy'. By about 1.00 a.m., the *Advance* was starting to drag anchor, so Breeze fired up his engines and stayed out of trouble by steaming into the gale all through the night. The only damage his ship suffered was a damaged propeller. He was in luck; most of the other craft that stayed in the vicinity were, in Carl Allridge's words, 'sunk or driven ashore'.

Not so fortunate was the fourth patrol boat, commanded by Lieutenant Bob Dagworthy. As the HMAS *Arrow* was riding out the storm, one of the ship's company remembers counting some seventeen vessels—prawn boats, two ferries and a schooner—anchored behind them in Darwin's inner harbour before the tropical cyclone struck. Soon after the 'hard blow' at around 2.00 a.m., the same crew member looked around and there was nothing left of these craft. All had presumably sunk.

The *Arrow*'s mooring line and 160-metre anchor chain partly snapped. While it was still half attached, Lieutenant Dagworthy started the engines. The patrol boat was rolling very violently. Then the chain snapped, and the windlass disintegrated as it was torn out of the vessel. Unable to gain headway, the skipper decided to run the boat up on to the wharf so that the nineteen crew could jump off and not be in danger of going down with the ship. The *Arrow* was driven out of control by the gale into the outer corner of Stokes Hill Wharf. It was rolling heavily in the crashing waves as members of her ship's company jumped on to the oyster-and coral-encrusted wharf. In absolute darkness, they had to cling for their lives to whatever they could. Petty Officer Les Catton was struck by flying cargo, knocked unconscious, and blown back into the water, where he drowned. Another sailor, John Rennie, suffered a similar fate. The men who did manage to clamber on to the wharf before the patrol boat sank beneath them had their clothes ripped from their bodies by the wind and driving rain. Four of the ship's company were swept into the water against the wharf pylons and struts; they managed to survive, but were severely lacerated in the process.

Those men who managed to get on to the wharf decking faced a new threat after nearly drowning. Steel cargo containers, some 'the size of a room', were being swept along the wharf and 'tossed along like children's toy blocks', according to one survivor. Once ashore and off the wharf, the sailors made their way to a deserted car park where some of the vehicles—including the assistant harbour master, Colin Woods' car—were blown on to their sides. Some other sailors took refuge wherever they could to avoid the flying objects, broken glass and other lethal missiles being blown before the storm. Skipper of the *Arrow*, Bob Dagworthy was unable to jump off his sinking boat and took to a life raft as his patrol boat went down. He was then at the mercy of the lashing winds and was found thirteen hours later floating on a life raft.

Perhaps the luckiest man afloat was one of the youngest skippers in the Northern Research (N.R.) fleet. Bob Hedditch, twenty-four, of Portland, Victoria, put out to sea with two crew aboard his 25-metre steel-hulled trawler, the N.R. *Anson*, at 7.30 p.m. on Christmas Eve when the tropical cyclone warnings were issued. 'At midnight it hit us,' he said. 'The wind blew in our windows on the bridge and tore the back door off.' Waves crashed into the wheelhouse, and the young skipper had to lie on the floor to steer; the gale blasting through the broken windows was too strong to allow him to stand. The winds, which reached 280 kilometres an hour, ripped off two booms at the stern and buckled steel guardrails.

Bob Hedditch kept his engine full ahead into the tropical cyclone, put out both anchors and hoped for the best. 'By 2.00 a.m. we had no lights, no steering and only the main engine to keep us going.' He said afterwards, 'We lost both our anchors and I didn't have a clue where we were.' At 5.00 a.m., *Anson*'s crew saw waves crashing on the rocks at Larrakeyah cliffs. 'We saw two distress signals—but there was nothing we could do.'

6 LANDFALL

SHORTLY AFTER MIDNIGHT on Christmas Day, the cyclonic gale-force winds intensified and the barometric pressure at the centre of the storm plummeted. When the storm struck the coastline at around 1.00 a.m., it was with a ferocity that few people could have anticipated. First to experience the brunt of the cyclone were the port and harbour area, the CBD, Larrakeyah, Stuart Park, Parap and Fannie Bay. The RAAF base and Ludmilla were next, followed by the Nightcliff and Rapid Creek areas. The wind's impact was unevenly distributed; some houses were sheltered by small hills or were in an area where other geographical features deflected the wind.

```
TOP PRIORITY. CYCLONE WARNING NO. 20 ISSUED BY THE
DARWIN TROPICAL CYCLONE WARNING CENTRE AT 1.00 A.M.
CST, 25 DECEMBER 1974.
AT 12.50 A.M. SEVERE TROPICAL CYCLONE TRACY WAS CENTRED
27 KILOMETRES WEST-NORTHWEST OF DARWIN AND MOVING EAST-
SOUTHEAST AT 6 KILOMETRES PER HOUR.
THE CENTRE IS EXPECTED TO CROSS THE COAST IN THE
VICINITY OF DARWIN AT ABOUT 5.00 A.M. TODAY.
VERY DESTRUCTIVE WINDS OF 120 KILOMETRES PER HOUR WITH
GUSTS TO 150 KILOMETRES PER HOUR ARE EXPECTED TO CONTINUE
BETWEEN GROSE ISLAND AND THE VERNON ISLANDS TODAY. TIDES
SHOULD NOT REACH HIGHER THAN HIGH WATER MARK.
```

Tropical Cyclone *Tracy* lumbered in across Shoal Bay bringing with her enormous amounts of wind-driven rain. The rainfall in

Darwin increased substantially after 2.30 a.m., with approximately 105 millimetres falling in an hour just before 3.30 a.m., before abruptly decreasing in intensity. It was later estimated that 280 millimetres of rain fell on Darwin and its environs during the twenty-four hours from 9.00 a.m. on Christmas Eve. The pressure gradient was almost vertical as *Tracy*'s barometric pressure dropped to at least 955 millibars. From the available barograph traces, the pressure gradients appeared to average about 1 millibar per kilometre over the storm. Near the centre, gradients of about 3 millibars per kilometre occurred. As far as tropical cyclones go, *Tracy* was relatively small, being only 40 kilometres across, but the severity of the winds produced by the extremely low barometric pressure made it a deadly and destructive storm.

Peter McIver, an apprentice machinist and speedway fan, had spent the evening at the Berrimah pub. He had had a big night, and had gone straight to bed after arriving home at Karinya Flats, in Mitchell Street, a major thoroughfare in the CBD, where he lived with his mother. When the storm struck the town centre, Peter was woken from a sound sleep by his mother. He couldn't believe the noise the wind was making. Luckily, it was screaming past their flat and not coming directly at them. This was the 'first wind', a precursor of the 'second wind' which would follow the passage of the eye of the cyclone through the town. From a window of the flat, Peter watched as two cars, which had been flipped upside down by the force of the wind, were pushed along the road. Through the lightning flashes, he could see the trees being bent over and stripped of their foliage. Most frightening for Peter and his mother was the incredible level of noise. 'It was screeching. My mum was crying and very upset. There was lots of lightning.' Peter stayed at the window watching the tempest blowing debris past their flat. He thought that the eye passed close by, but it was only for five minutes or so; and when the second wind arrived from the south, the noise was twice as loud. Fortunately, there was another three-storey block of flats behind theirs, which protected Karinya Flats from any major damage.

Around the corner in Shepherd Street, Eddie Quong and his family were doing it tough in their high-set wooden weatherboard house. Eddie recalled that he was trying to talk to someone on the telephone when the upstairs windows blew in. The noise level increased dramatically and he couldn't hear his wife Greta unless she cupped her hands over his ear. Eddie went downstairs to 'tell my two boys sheltering in an adjoining flat beside the house to barricade the door and not to move'. On his way back up the stairs, he said, 'the wind was so strong it nearly blew me off the stairs, even though I was hanging on with both hands and had an arm wrapped around the railing'. By the time the first wind had pushed through their area—a period of about one-and-a-half hours, according to the Quongs—they had lost all of their windows, and the back bedrooms had been wrecked when the next-door neighbour's roof was torn off and smashed into the back of Eddie and Greta's house. The impact had exposed the roof, and when the rain drove into their plaster ceiling, it soon became soaked and then caved in. All the new carpets were sodden.

About a kilometre away, at the southern end of the CBD, was the port and wharf precinct. The prawn trawlers were having the worst of it. Their high sides on the aft working deck gave the wind purchase, and those trawlers that broached and sank flipped over very quickly, giving their crews very little time to escape. Most of the craft in the inner harbour were pushed around, their moorings straining far beyond their normal capacity.

Geoff Crane and his family were at the Weather Bureau, where Geoff was back on shift. The operational shift was now eight people: a senior meteorologist, two or three other mets, two communications staff (telex) and two satellite imagery staff. Geoff's wife Barbara and the kids were on the same floor, 'out of the way' in a corner. When the first wind blew, the windows on the eighth floor flexed and water started to come in. There were several families who had decided to shelter in the building, and they all moved to the central stairwell, away from the edges of the building. They couldn't hear much noise and felt relatively safe so far up above the city. As it turned out, they were above the flying debris, which generally only reached four or five floors above ground level. Only one window, near the ladies' toilet, was punched in. As Geoff cleaned up the mess, he thought a bird may have been blown into the window. He wondered what a bird was doing flying around in those conditions. Barbara Crane felt concerned, but was grateful for the moral support of the other families. She felt for the mets on duty whose families were out in the suburbs and having to deal with the storm without them.

At around 1.30 a.m., Darwin was blacked out when the power failed at the power station. Only one beacon glowed in the CBD, the Weather Bureau, which had an emergency generator. Geoff recalls, 'It was the lonely light in town. People came to it like a moth to a flame.' The Cranes were unable to see out and had no idea of the devastation below at street level. It became very hot and muggy in the building, as the emergency generator didn't power the air conditioning. The MLC Building remained in the zone of maximum winds throughout the time Cyclone *Tracy* passed over the CBD. In the early morning, the Bureau's communications system failed. By 5.00 a.m., however, Geoff was able to phone the senior met in Perth, Steve West. He then phoned his parents in Victoria to tell them that they were okay. His parents had been unaware that a tropical cyclone had just scored a direct hit on Darwin.

In Westralia Street, in the coastal suburb of Stuart Park, Greg and Barbara Huddy had spent the evening revelling in the joy of their newborn child. Now they were battling to stay alive. They had gone to bed at 10.00 p.m. after unsuccessfully trying to soak up all the rainwater being driven into the house. Barbara recalled, 'I got scared after we went to bed. The house was rocking, and there were enormous amounts of thunder and lightning. It was truly frightening.' Their children were in a separate bedroom, and they had their newborn with them. Greg recalled, 'The silliest thing we did was go to bed in our pyjamas, and we weren't wearing any shoes.' When their glass louvres and windows blew in, there were shards of glass right throughout the house. When they decided to move everyone into the bedroom closest to the lounge, they all sustained cuts to their feet. The wind was roaring in from the north-west, but they were protected by a rise in the ground and suffered very little damage from the first wind, which Barbara estimated lasted about three hours. 'If it wasn't three hours, it certainly seemed like it,' she said. When the eye passed over the coast, the family moved into the bathroom, taking their mattresses with them for added protection against the wind that was now coming straight up from Frances Bay and smashing into the house, rocking it even more violently.

When the wind started peeling the roofs off houses, it allowed the wind and rain in to totally soak everything. Very often, the plaster ceilings then collapsed. Some lives were saved and further injuries prevented because people were being cocooned under the rubble. More devastating was the impact of the wind on the interior walls, which were placed under enormous stress. If the wind got a purchase on the walls, rooms on the ends of buildings—which tend to be structurally the weakest—were blown away. The resulting debris then smashed into other houses, weakening their structure and sometimes ripping off their roofs, creating a snowball effect.

Sheltering in their house in Stuart Park's Gothenburg Crescent, not far from the Huddys, were Marshall and Cherry Perron. At around midnight, the couple decided to move from their bedroom, which was taking the brunt of the strong winds that were pummelling the front of the house. 'The noise of the house shaking, windows rattling and debris smashing into our house gave us cause for great concern,' Cherry recalled. The Perrons decided to move into the spare room, which was furthest from the danger. A 'sound like a shotgun' rent the air; the large sliding plate-glass doors in the lounge room had bowed to their limit, and when a severe gust slammed into the house they blew in, showering glass splinters over about 10 square metres of lounge room and hallway. Marshall described it as 'a huge explosion'. The wind was now roaring into the lounge room area, creating an enormous and instantaneous area of very high pressure under the roof. Marshall continued, 'The wind then took off the roof, and all we were left with was the ceiling. Rain then came in through the plaster. The whole roof was gone.' Marshall and Cherry—and their very frightened cat—sought shelter at the opposite end of the house. They had a mattress with them, and pushed a single bed up against the inward-opening bedroom door.

Outside, across Frances Bay, they could see enormous sheets of lightning. The noise level rose sharply once the roof went, and thunder exploded all around them. But what was really worrying now was that they could hear the sound of debris and swirling glass—as if it was in a huge blender—coming from the hallway. They were now more or less trapped in their own refuge and didn't dare open the door for fear of being shredded by the churning glass. The main bedroom windows had broken and added to the large amount of glass in the hallway. Not to let the storm put a damper on things, Marshall looked at Cherry and said in a deadpan voice, 'It needed a paint anyway.' The kitchen cupboards had their doors ripped off and were emptying their contents of crockery and saucepans into this internal maelstrom. Cherry commented that it sounded like the proverbial 'bull in a china shop'. The Perrons' greatest concern now was that the windows in the spare room, where they were sheltering, might blow in and they would then be at the mercy of the storm. All they had left was a single bed, which wasn't going to offer them much protection, and the built-in wardrobe, which wouldn't easily accommodate Marshall's height. Listening to the roar of the wind, the thunder and the glass blender outside the bedroom door, Marshall reassured Cherry, 'As long as we are physically okay, nothing else matters.' The greater part of their assets were gone with the wind, but they felt reasonably safe 'provided it stayed as it was until it died down'.

Laurie Coffey's family, who had hoped that this year Laurie's work on the barge *Allana Fay* would allow him to join them for Christmas, were being pummelled by the storm in their house in Larrakeyah. Laurie's twelve-year-old son Matthew had sought shelter with his mother, sister-in-law, younger brother and two sisters in their 3 x 5 metre bathroom. Matthew and his brother suffered damage to their ear drums when they were caught in a vacuum in the almost airtight room. Matthew could feel his ears constantly 'popping'. After moving out of the bathroom, everyone sheltered in the hallway, and then a tree fell on the house. As the storm raged around them, flying debris all but demolished the superstructure of the high-set house, apart from the hallway where they were sheltering. The ceilings caved in, and they were extremely wet and cold. They could see the pavlovas that Matthew's mother had made for Christmas lunch floating around in the dining room. A large piece of masonite board then fell over the huddling group, giving them some protection during the rest of the storm, though they were pelted constantly by horizontal rain. The house rocked 'like a show ride', Matthew recalled.

When the eye of the storm came over, Matthew noticed a silvery-grey glow in the sky. 'It was dead silent,' he recalled. 'Apart from dogs barking and next-door's chooks cackling.' When the second wind came with a deafening roar from the south-east, the family rode it out where they were. Their house was almost totally wrecked.

At Larrakeyah Army Barracks, Captain Peter Kerntke was in the operations room. He had lost his strategic communications link to the RAAF base when the antenna had blown away; the UHF telegraph was also down. Peter knew it was essential that they maintain communications in order to enable quick recovery action after the storm passed. He sent one of his radio operators, Doug Purcell, in a Landrover that was fitted for radio (FFR) over to the RAAF base to re-establish communications. In the ops room, things then started going haywire, Peter recalled.

 The telex started randomly hitting up [coming on line and making scrambled messages] and the normal telephones also went. I knew the telex was going to go. I told my signals supervisor, Tanker Hawken, to establish a stand-by link to 104 Signal Squadron in Townsville on HF [high-frequency very long range radio] and have them on listening watch to quickly re-establish communications.

Much to Peter Kerntke's chagrin, that order was countermanded by the commanding officer of his headquarters. That officer apparently didn't want soldiers (who were rostered on duty) to have extra work on Christmas Eve. Peter still had radio communications with his street wardens through backpack VHF radios and, with Captain Barry Darr, was coordinating the movement of all families into the Messes. At 11.45 p.m. he asked Captain Ian Stewart to collect his wife Diane and their children from their married quarter on Nurses Walk and take them to the shelter in the Officers' Mess. Peter explained, 'It survived World War Two, so *Tracy* would be no sweat.' Diane Kerntke and her children were taken to the Officers' Mess, but not without some drama. When she was trying to get down the stairs of their married quarter, the wind ripping up off the cliffs of Doctors Gully was so strong that she couldn't carry the children or the suitcase she had packed, and had to hang on to the railings. She was helped down by Captain Ian Stewart who bundled her and the children into a car and drove them to the shelter.

Across the road from the Kerntkes' married quarter, Captain Denis Castle saw the entire roof of the Kerntkes' house disappear only fifteen minutes after Diane and the children had left to go to the Mess. 'Then the whole house was destroyed,' he said. At the Officers' Mess, most people were upstairs until the windows blew in. When a woman was hit with flying glass, everyone relocated to the downstairs dining area, which had fewer windows and was much less exposed, where they sheltered under the fairly substantial 3-metre-long dining-room tables. When the power went off in the Mess at 12.30 a.m., they were huddled in the dark listening to the roaring tempest outside. Diane, who had been soaked while getting herself and the children out of the house, cradled them in her lap as she shivered uncontrollably.

Not long after midnight the barracks lost all communication links, including the Landrover FFR that had been dispatched to the RAAF base. Peter Kerntke:

 Soon after 1.00 a.m., I told Doug to come back [to Larrakeyah] in the FFR. He came back into the ops room looking like a mad cat. He said that on the wild drive back to Larrakeyah, he had been in neutral but was doing 30 kilometres per hour with a following wind!

Peter remembered that he and Diane had been asked to look after their neighbours' dogs, so he jumped into a Landrover and went to collect the frightened animals. While he was hopping around trying to gather up the dogs, the plaster on his broken leg got wet and soggy and eventually fell off.

Back in the ops room, the men on duty waited until the storm hit and then tried to get things ready so that they would be up and running again as soon as it had passed. At one stage during the first wind, Peter looked across at the ops room's Besser-brick wall, which faced the west.

> It was moving in the wind, and that howling wind went for a long time. I really thought that the wall would collapse. It had to be waving 30 centimetres at the top. I seemed to be always going to the toilet—lots of nervous pees, I guess. The water level in the toilets was low. I went out and checked the emergency generator that had kicked in. The water in the generator room was 30 centimetres deep. There were big iron doors to that generator room and they had blown off—disappeared. So I locked the access door so that people wouldn't get electrocuted if they stood in the water.

Peter thought that the eye came over the barracks at about 1.30 a.m. and that the wind died completely at around 4.00 a.m. By then, Captain Peter Kerntke, communications officer, had no portfolio—he had no communications of any kind outside of Darwin and was rueing his commanding officer's decision not to have a link to Townsville by Morse code. The only thing he could do was re-establish communications within his own area, and he had six FFR vehicles to enable him to do that. But first his men would have to replace all the smashed antennae to allow the radios to get some range.

About a kilometre away, Julie Tammo was awake, but her firefighter husband Heikki was sleeping like a log despite the maelstrom. Heikki had gone to bed early, as he was rostered on duty on Christmas Day. Julie ran around the house trying to soak up the water that was being pushed in through the closed louvres. Like many people in Darwin that night, she thought she was the only one whose house was being inundated with water or whose roof was being peeled off.

There was much cursing of builders, plumbers and carpenters among the besieged residents of the stricken town, who had no idea of the extent of the devastation *Tracy* was causing.

The Tammos had lost power quite early and Julie was working by torchlight during the first wind. Finally, she hauled her husband out of bed when the noise level had become a roar, and that was when their roof started to go. Julie, who still thought the damage was isolated, asked Heikki to ring the police to get help with the roof, but the phone was dead.

Grant and Sandy Tambling were in Kahlan, the suburb just north of Larrakeyah. Fortunately, their architect-designed, four-year-old house was behind a hill, which offered some protection from the wind. Grant had been tossing up whether to vacate the house because, although they were in the lee of Larrakeyah Hill, they were down low near Kahlan Oval and were looking across at the ocean not more than 500 metres distant. It was dead flat between their house and the water. But for the moment, the wind was sparing them the grief that many others were experiencing. The wind was coming mainly from the north and deflecting around behind their house. When their dinner guests had departed, Grant had taped their large picture windows and filled the bath with water. They didn't change out of their clothes, and had no thought of going to bed. Their two very young children were asleep in their beds, but when the wind slammed into the rear of the house and debris started smashing into the walls, the whole family relocated into a corner of the master bedroom. Grant and Sandy gathered all of their mattresses and made a protective bunker that they then placed over everyone.

The house started taking numerous hits from flying debris, but thankfully most of it was small stuff. Grant was amazed by how far glass could bend before it would break. They had large windows in aluminium frames, and the glass was bending an extraordinary amount and almost falling out of the frames. The wind pressure was so great that the frames opened and the curtains were sucked out through the gap; later the frames snapped back, but the curtains were still outside the windows. Grant was worried that the glass windows would also be sucked out, which would further lessen the integrity of the house. Occasionally, during brilliant flashes of lightning, they could see across the oval. Grant doesn't think that the eye of the cyclone passed over their house, but he recalled a small lull when the wind decreased a little and then changed direction. His greatest concern was where they would go if the house started to disintegrate, because they had no concrete block laundry under their house, just their cars. The other problem was that if they *did* have to evacuate, they would have to go uphill and into the wind. People had been advised not to shelter in their cars for the very reason that young Peter McIver had witnessed over in his street. If the glass is broken, the occupants have little or no protection. Or if the car is under the house, it may be crushed if the house collapses or a heavy item such as a refrigerator falls through a weakened floor.

The second wind hit the Tamblings' house with more ferocity than the first and was now coming from the south-east. Grant recalled:

> We suffered more structural damage this time, but compared to most other people in Darwin it was minimal. There was lots of penetration to the roof, and the fibro walls were punctured and fractured. We had a fair bit of water damage, but we still had our house. I believe we were among the 10 per cent of the population to retain their house.

Remarkably, the Tamblings' above-ground pool remained intact. However, if the tide had been high, they could have had a problem with storm surge.

Behind Kahlan and another kilometre away, Nick Paspaley was in his parents' house at Myilly Point. The exposed house was suffering greatly at the mercy of the roaring winds, and they had to move from room to room. Nick recalled:

> By the time the cyclone had finished, the house was destroyed and we ended up huddled in the cellar downstairs, including the old man. I went upstairs to get him. I went down to his room and said, 'Come on, Dad. Let's go downstairs. It's getting serious.' He thought so too, and he got up. We were going through the lounge room and there was a big window . . . It was like slow motion—it just disintegrated and was blown in towards us. Dad made it past the door just seconds before the glass burst through into the room.

A house typical of Darwin before Cyclone *Tracy*—high set with a storeroom and laundry area underneath, and surrounded by lush tropical gardens. After *Tracy* it took several years for most of the foliage to fully regenerate. (COURTESY JOHN AND HELEN AULD)

Rows of houses were cleared of their entire upper structures as debris snowballed along the street. Very few dwellings were left without major structural damage. (COURTESY WERNER JAMNIK)

After *Tracy* this house had nothing left except its steel floor bearers and a part of its toilet. Several people were killed when large objects like refrigerators crashed through weakened floors. (COURTESY ALAN GROVE)

The debris that littered the yards made it extremely hazardous for people to move away from their dwellings. Many were killed or injured by flying corrugated iron as they tried to escape the wrath of the storm. (COURTESY AUSTRALIAN ASSOCIATED PRESS)

The worst damage to Darwin was in the northern suburbs. Here, most houses were totally destroyed, as the flat ground offered no protection from the furious onslaught of 250-kilometres-per-hour winds. This was the scene on Christmas Day not long after first light—an unbelievable picture of utter devastation set against threatening and rain-heavy skies. (COURTESY WERNER JAMNIK)

Even steel power poles were bent and twisted by the fierce winds. Very few of these poles survived the onslaught of *Tracy*—most were rendered unusable. (COURTESY WERNER JAMNIK)

ABOVE: Danny McIver stands forlornly among the wreckage of his home holding his two remaining possessions—his cat and his bathroom sink. (COURTESY FAIRFAX)

LEFT: Looking towards the Casuarina Square shopping complex with Fannie Bay in the background. In the northern suburbs, little shelter remained. (COURTESY ALAN GROVE)

RIGHT: A light aircraft after it had tumbled along the runway during the storm and smashed into a hangar. Not one aircraft in the open survived the cyclone's fury. (COURTESY FAIRFAX)

In the storm, the floorboards of this house were flipped over the top of the structure. The street sign in the foreground was ripped out. Many residents became lost when all recognisable landmarks were taken away by the winds. (COURTESY ALAN GROVE)

ABOVE: A family hangs out their drenched belongings on Boxing Day in an effort to create some order out of the chaos. Very few trees were left with any more than their larger branches after the cyclone had passed. Birds and small native animals were not seen for months. (COURTESY FAIRFAX)

RIGHT: The Royal Australian Navy's patrol boat HMAS *Attack* lying on the beach at Doctor's Gully. The boat was driven ashore dragging a 7.5-tonne concrete mooring block along the harbour bottom. In the background, the badly damaged Iron Ore wharf can be seen—several vessels smashed into the wharf and sank.
(COURTESY *NAVY NEWS* RAN 2663/1)

LEFT: Cars in the swimming pool of the Travelodge Hotel in Darwin City. Many cars were pushed along the roads with terrified residents trapped inside. (COURTESY FAIRFAX)

BELOW: The backyard of John and Janice Woodcock's house, with John's parents' caravan in the foreground. His parents lost all their worldly possessions when the wind tore a hole in the roof and sucked out all of the van's contents. The Woodcocks spent the storm out in the elements after their roof disappeared and their ceiling collapsed. (COURTESY JOHN AND JANICE WOODCOCK)

Having a shower was difficult, but Laurie Gwynne found a solution at Berrimah research station about two days after the storm. Laurie had all of his toes broken when he was hit by flying corrugated iron. (COURTESY LAURIE AND SHIRLEY GWYNNE)

The devastated city of Darwin was unable to support more than a quarter of its population, and so a mass-evacuation began on Boxing Day and lasted until New Year's Eve. Almost 20,000 people were moved interstate via air in the six days. Another 6000 left by road. Many survivors never returned to Darwin. (COURTESY FAIRFAX)

Although the suburb of Ludmilla is a little further inland, some freaks of wind hit in isolated areas such as Peter Coombe's Street. Peter and his wife Jenny, who was pregnant with their first child, lived in Wells Street. Peter had taped the windows on their house, which he described as a 'typical Northern Territory Administration high-set, concrete pylons, iron roof, asbestos cement-cladding walls and with a storeroom underneath. It had established gardens and was very tropical.' The Coombes had had thirty people around for a Christmas Eve party. At around 9.00 p.m., trees started falling on to houses and a couple of hours later the power went off. It was a bit difficult to have a party in the dark, so the 'party animals' decided on a change of venue. Peter Coombe is a fairly laid-back sort of character with a dry—if not arid—sense of humour. He recounted what happened next:

> We decided to drive down to East Point [about 5 kilometres distant] to view the cyclone. We went in convoy, but as we got to Fannie Bay [half-way] a tree came down across the road and hit the leading car, so we returned home.

For a brief time the party continued upstairs, but people soon dispersed back to their own homes—the Pickerings, who lived just around the corner, were among them. Peter and Jenny then went to bed. Their house was partially protected by a hill that deflected the wind. But more-exposed houses across the road were being slowly destroyed. The wind was screaming, and heavy rain was pelting against the windows, so Peter and Jenny were 'not really asleep'. Peter could hear windows smashing elsewhere in the house. When they shattered, he said, 'the noise was incredible'. What happened next was bizarre. Peter Coombe told the story as if this sort of thing happened every day:

> Three people and a baby—who were strangers—came into our bedroom. It may have been around 1.30 a.m. They had broken down our front door and were hysterical. One of the women had gashed her arm quite severely when

she opened the front door through the broken glass in the door window. We were lying in bed, and then lightning flashed and we saw them. They were real panicky, so I put them in our bathroom. I then took the injured woman to hospital.

Driving along roads where it was extremely difficult to see probably wasn't a good idea, but the woman was bleeding profusely, Peter explained, and needed urgent medical attention. Her husband was totally incapable of handling the emergency. So, Peter—wearing just his underwear—and a hysterical bleeding woman climbed into his Peugeot sedan and headed off into town. It was raining so heavily that Peter had to drive at a crawl and follow the white line in the middle of the road—it was the only reference point he had to keep him on the road. As he said, 'The rain was horizontal and it was pitch black.' When he was near the Motor Registry building on Goyder Road, about 2 kilometres from home, a startling thing happened.

> I was driving along and all of a sudden I hit a house. It was right in the middle of the road. I knew that it was, because I was following the white line. It was a complete house still with its roof on and fairly intact. It still had its banks of louvres in place. I wasn't going too fast, so it wasn't much of an impact. I backed up and carefully drove around the house and continued on to the hospital.

When Peter and his injured passenger arrived at the casualty ward at Darwin Hospital, some ambulance officers complained when he parked in a 'no parking' area reserved for emergencies. He ignored them and took the bleeding woman into the ward. He then headed back home to Ludmilla—another 6 kilometre drive in raging winds and rain. His car was being buffeted severely, and Peter admitted that he was lucky he wasn't hit by anything. When he came to the spot on Goyder Road where he had run into the house, 'the house was gone'. The only damage he suffered during this incredible journey through gale-force winds was when a coconut went through the window of the car. If it had hit him, it would have killed him instantly. When he returned home, he decided to relocate to a friend's house like his good friends, the Pickerings, had done.

Judy and Tony Pickering and their young son Ben arrived at Kelly Morrissey's house in Clancy Street, Fannie Bay. They left their pet dog, Harry, a 'labradoodle' cross, in the bathroom with a bathtub of water. They had been forced to motor down the road at a crawl because of wind, rain and almost zero visibility. All along the way, small trees were being ripped out of the sodden ground and thrown across the road. When they arrived at Clancy Street, Tony noticed that a huge rain tree was down across the pool. They went upstairs initially, but soon everyone at the house moved downstairs into the very sturdy brick rumpus room. They all had to hold hands and 'daisy chain' down the stairs. 'The ferocity of the wind was exceptional,' Judy said.

Around twelve people were now sheltering in the 12-square-metre rumpus room, which included a laundry. Adjoining the room was a toilet and shower. There were large, sliding glass doors that looked out over the swimming pool. The group had gathered all of the mattresses and packed them against the windows. The party that had started at Peter Coombes' had now relocated. They had plenty of food that had been brought across from the party, and they now waited for the storm to hit. The first wind wasn't a blast, but it slowly increased. The group looked nervously at each other as they heard debris slamming into the outer walls and the superstructure of the house above them. Judy Pickering covered her son Ben with her body as she lay down on the carpet.

After an hour or so, the superstructure above the group started to disintegrate off the floorboards. More and more debris was smashing into the house. The screeching, tearing noise of corrugated iron pierced the black night. The men in the group were holding the mattresses against the sliding glass doors to stop the windows from caving in. Like many glass objects that night, they bent well beyond reason—but held in place. Above them, the walls were pulling off the floorboards. Judy Pickering explained, 'Now it was really scary. Dirt like fine sand was coming through the floorboards.' The wind and rain were forcing the dirt under the carpets and between the tongue and groove flooring like fine mud, and it fell on to the women huddling in the room below. Later, Judy and her son would develop conjunctivitis. The screaming and howling of the wind seemed interminable. At one stage, Judy cried, 'Please God, let it stop.' The men bracing against the glass doors were concerned about what would happen if they couldn't hold on, as the wind was now coming straight at the doors. The first wind lasted until around 3.00 a.m. Peter Coombe was helping to hold the mattresses against the doors and could see out into the yard where the rain tree was down over the pool. He commented on the amount of lightning. 'It was surreal—stroboscopic, in fact. One minute it was almost like day, then the next it would be pitch black.' He watched in fascination as the rain tree was blown across the pool in the first wind and then back from whence it came with the second.

Back at Porter Street in Ludmilla, the wind was so noisy that thirty-three-year-old engineer and earth-moving contractor Alan Grove and his wife Joy couldn't sleep. They had put their three children, aged five, three and four months, in one room and, unexpectedly, they had slept through the first part of the storm. The first wind had been building and reached its peak after midnight. The wind was screeching, and the sky was dark with heavy, horizontal rain. But the wind was coming across a sports oval, so there wasn't too much debris hitting the house at this time. Alan recalled that the 'glass louvres were bending about 60 millimetres and not breaking'. There was a lot of water over the floor, and he and Joy lifted the matting they had on the wooden floorboards and let the water run out through a knot-hole. It was a worrying and, as Alan described it, 'lonely time. We had no power, no radio and no knowledge of how long it was going to last.'

The second wind hit the Groves' high-set, boomerang-shaped house so violently that, in Alan's estimation, 'it was rocking 100 millimetres off the floor joists'. Now everyone was wet and very cold. Alan remembered having a lot of 'nervous pisses'. They hung on; the house was still holding.

Across the main arterial road of Bagot Road at the RAAF base, two families were toughing it out. Dianne Ferguson, her husband Bob and their two sons were sheltering in the bathroom after they moved the boys out of the bedroom when the manhole cover was sucked up into the ceiling and disappeared. The house was a high-

set married quarter, but the bathroom stuck out from the rectangular shape of the house and was exposed. Dianne recalls that when
the first wind reached its peak after midnight, she felt 'at the mercy
of the elements. The wind sounded like a test run for a 747 Jumbo
jet, maybe ten of them all at once, overhead.' It was truly frightening, she said, and the family huddled together in the small
bathroom. The sound of sheet metal and corrugated iron banging
and clanging as it wrapped around poles and fences was deafening.
Dianne thought that their front stairs had been ripped loose and
'were flapping in the breeze'. The family was all sitting on foam
rubber on the floor. Bob had nailed a blanket over the bathroom
window to stop debris and broken glass from flying through and
cutting them. The tension was relieved for a moment when their
youngest son, Kerry, asked if he could have a bath. Their four-year-
old son Michael was becoming very frightened, and so everyone
sang Christmas carols to take their minds off the storm. Bob was
having the worst of it, because every time the house shuddered
under the impact of more debris smashing into it, Dianne—who
was clinging to him—would pull more hairs from his legs when she
flinched.

   Just down the road, less than 500 metres away, Josephine
Foreman had gone to bed. Her drunken husband was asleep beside
her and she was listening to Midnight Mass on a small transistor
radio. Near the end of Mass, she heard storm noises, smashing glass
and 'tinkling' sounds being picked up by the microphones at the
church service. She was sure she heard the minister tell the congre-
gation to 'get home now'. She decided to get out of bed, as the wind
was now roaring and she was worried about her kids, especially her
crippled son, Martin. The house was creaking a lot, so Josephine
decided it was time to get everyone into the bathroom. As she was
preparing to do that, the married quarter behind her house, which
had been protecting them (to an extent) from the wind, left its
pylons and smashed into their house. Most of their windows were
broken, and the main bedroom was penetrated with timber and iron.
The kids' bedroom windows were smashed in as well, and when
Josephine went in to collect Martin she found he was facing
inward, away from the window, and was covered in broken glass.

Her other son, Dwight, had waist-deep glass in his bed. Luckily the children had sheets over them, and Josephine peeled back the sheets and glass when she picked them up. Walking barefoot over the broken glass, she carried Martin down to the bathroom. She then went back for Dwight. Her husband grabbed a mattress for protection in the bathroom, where they sat on the floor of the shower recess. Josephine now had her family together in one place.

The wind pressure became tremendous, and Josephine felt that they were being sucked towards the door. She recalls that 'the noise was like a train coming at us. It wouldn't stop. We could hear tin peeling and rolling down along the road.' Her husband moved out of the shower recess and pushed his back up against the door because it looked like it was going to go. She said, 'The house rocked like on an old train, and I was worried that the floor would go next. We sang and I told stories.' Josephine estimated that the first wind lasted around two hours. 'I really thought that we might die.'

Across the other side of the main airport runway, two of Bill Gough's men were fighting a battle of their own with the aircraft that couldn't be flown out to Katherine that afternoon. The hangar doors had been ripped off, and when the wind direction changed the men moved the aircraft around by pushing on its wheels. Their aim was to give it as much protection as they could by placing it out of the direct blast of the wind inside the hangar. They would continue to do it all night—and saved the plane, the only one at the airport that survived.

The suburb of Fannie Bay, as its name implies, faces directly on to the sea and the gale-force winds smashed into the suburb with venom. Carl Allridge, the harbour master, was with his wife, Joan, and their two teenage daughters at their house in Parsons Street. They sheltered in the lounge room with their backs to the storm. After the first wind really got wound up, a friend and neighbour and her four children and a menagerie of cats joined them. Carl recalled that 'trees were going over, the power was off and there was water everywhere inside the house'. The house survived the storm, although it was badly damaged. Carl went downstairs during the eye to have a look around and see what condition the underneath of the house was in, but when the noise of the approaching second wind started he dashed back upstairs. In the lounge room the glass in the windows was bending dangerously towards them, so they sought refuge under the kitchen table. The roof had been stripped of some iron, and the ceiling collapsed. Freezing rain drove in on the group and Carl felt that 'things had taken a turn for the worse'. The second wind was coming from the south-east quadrant. Carl said, 'The second wind was strongest. I now felt most at peril, because that was when the roof went and the glass blew in. The noise was much, much worse than the first wind.' They rode out the storm under the kitchen table—except for Carl (he couldn't fit), who sought shelter just to the side and was very, very wet and began shivering uncontrollably. He added, 'The killer was all the debris flying around.'

Only 100 metres away in Holtze Street, Fannie Bay, Elspeth Harvey had gone to bed early, at around 10.00 p.m., thinking the storm was just going to be a severe blow. Rain was starting to come in through the louvres, so she placed towels under the windows around the floor. She was in her pyjamas, and her 'puppy dogs' were up on the bed. Elspeth was angry at the rain because of what it was doing to her very nice house. There was a lot of water on the floors because the roof had gone, but Elspeth and her husband hadn't noticed it through the noise of the storm. The water, wind and rain got 'past the point of being just annoying' and so the family got dressed and, carrying the two dogs, went down the outside stairs to get in their brand-new Datsun 180B sedan. Elspeth's husband, Alex, who was wearing thongs, slipped on the wet front stairs and badly bruised his leg. Squeezed into this small car were three adults and two small dogs. They should also have had a cockatoo and the pre-school's budgies that Elspeth was looking after over the holidays. But when Elspeth went to retrieve the birds from under the house, their cages were gone. The Harveys took refuge in their car because the laundry area was open to the storm and their car was the last place left that was reasonably dry. But they were still very cold.

The first wind came from the sea and what Elspeth believed was the north-west. Everyone was soaked. At around 1.00 a.m., during the eye, Elspeth decided to return to the house to get a change of clothes for them all, but she couldn't get up the stairs, which were covered in rubble and debris. However, while she was out, she saw the budgies' cage in the backyard, with the birds still in it. She grabbed the cage and made her way through all the rubbish lying in the yard and back to the Datsun. The wind was sandblasting the car while they were huddling together. Elspeth recalled:

> My daughter Gail was okay, probably frightened, but was kept busy looking after all the animals. My husband Alex was very quiet. The car was being rocked violently. The cat was having a bad time. The pups were sleeping. My greatest concern was the cocky and where he was.

Milton Drew was alone in his company house just up the road in Wells Street and thinking how glad he was that his wife and family weren't having to go through the storm. It was a pre-Second World War, high-set house and was badly feeling the effects of the gale, especially the gusts that slammed into the house every minute or so. Milton recalled, 'It was really blowing by 10.00 p.m. and I opened up all the windows on the lee side. I was in the main bedroom attending to the windows when I could feel the pressure drop inside the house.' He sat it out in the end bedroom until about 11.00 p.m. During the first wind, the house suffered no significant damage and none of the windows had broken. He couldn't see outside, because it was pitch dark. When the second wind hit, he said, 'It was more powerful, at least ten times stronger. It blew all the windows in. I said to myself, "I'm outta here" and I went to the kitchen.' There were metal louvres in the kitchen and he thought he would be safe with them. Within thirty minutes all the louvres went, and glass showered all over the top of him and over the floor. He moved again to the hallway and realised he had very few options left. He was forced out of the hallway when some of the walls disappeared, so he moved next to the back wall of the house. He sat there on the floor all alone, and then his gas lamp went out. He was relatively safe in the room where he was, but then the remainder of the house started to disappear in bits around him.

Just over the crest of a nearby hill, a house behind Milton Drew's lost its entire roof. It flew over the top of Milton's house and landed on the street in front. Describing the noise of the impact, Milton said, 'It was like a jet going in. The house was shaking so much I couldn't lean against the wall. I started to wonder if I could survive.' The noise was starting to get to him. The roaring and screaming seemed never-ending. He had now lost about a third of his roof, and rainwater was teeming in on top of him. Milton recalled:

> It was endless; this incredible noise just went on and on. It was very scary, but I was grateful the kids weren't there. I didn't want to go into the bathroom, because the wall had been pierced by a 10 x 5-centimetre piece of lumber, which had penetrated about a metre into the bathroom. The scary part about being alone was the effect on yourself. I was no longer in control. I just wanted to look outside, and I was dying for the sun to come up. I wanted it to be just a bad dream. It affects you mentally.

The dangers of taking shelter in a car under the house were demonstrated most dramatically across the road from Milton. His neighbour, thinking his house was going to collapse, climbed into his car, which was beneath the house. He was crushed when the end of the building collapsed, trapping him inside.

Although often referred to as part of the northern suburbs, Nightcliff and Rapid Creek are also right on the ocean. The first wind was severe, but when the second wind came after the eye they got hit, and hit hard, as *Tracy*—still only moving forward at about 6 kilometres per hour—blew in across the Nightcliff Peninsula.

Chris Collins, his wife Trisha, and their cat and dog were sheltering in their lounge room. They had changed into storm gear—jeans, work boots and jackets—and were probably among the very few who were dressed adequately for the storm. They had no breakages during the first wind, though there had been increasing horizontal rain and, as Chris recalled, 'stuff bouncing off the roof, and leaves were being plastered against the windows'. The wind increased as *Tracy* moved over the peninsula, and they came under the influence of the more powerful winds near the eye. When the strongest gusts hit in the first wind, the power lines were ripped off the steel poles and then the power went off. Chris Collins recalled, 'It was well before the eye. The lightning was more like static discharges. There was so much wind noise I don't think we heard the thunder.' The eye passed over and lasted about twenty minutes. Chris is a commercial pilot who conducts aerial surveys and has seen the atmospheric phenomenon known as St Elmo's Fire. He described the sky as having 'an eerie green glow, rather than crashing lightning'.

Then came the second wind. 'It was like a thousand express trains coming at us,' Chris recalled. 'I had never heard a noise like it. It was far more destructive and had incredible strength.' Chris and Trisha quickly went into the bathroom, which had a shower, toilet, sink and a wall cabinet. They noticed 'more light coming into house'. The ceiling and the roof structure were lifting off the walls in the living room, which was now closest to the second wind. Trisha saw what was about to happen and cried, 'It's not fair, it's our house!' While they were in the bathroom, flying debris punctured the wall and Chris was hit in the back of the head and on his back. The dog and cat stayed close by, needing to be comforted.

Down the hall from the bathroom, the whole side wall of the house then collapsed. Chris was trying unsuccessfully to hold the bathroom door shut with his back. He was hit by bricks flying through the air 'from somewhere' nearby, though he recalled being 'just bruised and scratched'. Chris Collins is a tall, strong fellow, but despite putting his whole body into it he couldn't close the door owing to the pressure of the wind. Then he spied the 10-millimetre steel padbolt and slipped it into the door lock—it held! But their options were quickly running out, as they could hear their house being demolished. Chris and Trisha were now huddled together with the cat and the dog in the small shower alcove. They were being blasted by stinging rain, dirt and wind. There was continuous banging overhead, and eventually the Masonite board ceiling collapsed and drooped down, partially shielding them from the elements. Chris said, 'It saved our lives.' He recalled wondering, 'How much more of this is there? What are the odds against us surviving?' It was pitch black while they rode out the second wind and they were totally drenched. Soon they became very cold. Chris said, 'I was worried about hypothermia. We were in a big air-conditioner and just frozen. The cat was stuck to my chest.'

Young Nat Guinane was in Bougainvillea Street, Nightcliff. Her family was sheltering on the ground floor in a concrete block of four two-bedroom flats only 200 metres from the sea. When the first wind got up around midnight, a woman upstairs with two babies began screaming. The upstairs roof was going, the power was off and it was dark. The lightning increased dramatically and roofs were being blown everywhere. Nat recalled, 'It was a frightening noise, with what sounded like explosions, timber being ripped off floors, brickwork falling, and thunder.' She could hear corrugated iron being ripped off roofs and clanging and screeching along the street. The wind came in from the sea and took out their flat's windows. Her family then relocated to a flat immediately adjoining theirs, with seventeen other people. When Nat's father ran back to their flat to get some medication for his wife, he received cuts to his leg from the flying debris, but he wasn't too badly hurt. All of the people from the upstairs flats were now with this group, and they sheltered in the hallway as far from the wind as they could safely get. Some people soon became hysterical. Nat Guinane recalled:

The smell of garlic was unbelievable, as they had all just eaten a big meal. It was humid and hot in the cramped passageway. My greatest concern was that the concrete roof from the upstairs flats would fall in on us. The noise from upstairs was really bad, with bricks falling and smashing into the cars outside, and the sound of tearing metal and timber. I felt I was going to die.

7 OMEGA'S PATH

TROPICAL CYCLONE *TRACY* didn't follow a straight line when she made landfall. As had been her pattern for the previous four days, she moved in a general direction but rotated around and across a centre line of that directional track. This trochoidal motion continued when the storm hit the coast and the eye of the storm passed to the north of the CBD, then spun over towards Stuart Park and then up through the coastal and northern suburbs. The eye was 13–14 kilometres across and, due to the storm's trochoidal motion, it could be in one part of a suburb but not another. Those parts of the town that the eye missed experienced maximum winds for around five hours. The path of the cyclone was difficult to determine, but appears to have followed one similar in shape to the Greek letter *omega*, which looks like a horseshoe with the curve uppermost. When the cyclone passed over the western extremities of the suburbs, it headed off down the Stuart Highway towards Howard Springs and Humpty Doo and eventually petered out as a severe rain depression well south towards Katherine.

TOP PRIORITY. CYCLONE WARNING ISSUED BY THE DARWIN
TROPICAL CYCLONE WARNING CENTRE AT 12.30 A.M. CST,
25 DECEMBER 1974.
TROPICAL CYCLONE TRACY WAS LOCATED BY RADAR AT MIDNIGHT
30 KILOMETRES WEST-NORTHWEST OF DARWIN MOVING EAST-
SOUTHEAST AT 6 KILOMETRES PER HOUR.

The warnings were now almost irrelevant; the destructive winds were well and truly upon the people of Darwin, and the noise of the raging storm made it very difficult to hear the radio anyway. *Tracy* started smashing into the northern suburbs of Rapid Creek, Jingili, Alawa, Wagaman, Nakara and Wanguri. The suburbs were newer the further one went north. The gardens in the more recently established suburbs had less vegetation. The one distinct feature of the northern suburbs is that the ground was now quite flat, with no significant topographical features. There were no hills to provide protection or deflect the wind when it came.

Chris Kingston-Lee lived on Trower Road, Rapid Creek, which is the main road that leads to the northern suburbs. Some time after midnight, he heard things smashing. He dragged the double bed into the loungeroom and used it as cover for his wife, Roslyn, and one-year-old son, 'TJ'. Normally their house would be brimming with kids, as the Kingston-Lees were a Uniting Church foster home for children. But it was just the three of them. As debris started smashing into the house, the family huddled closer together. Chris' dog was right under his feet and clinging to him like a second skin. Their cat had gone; it took off and was never seen again. Chris guessed it was about 1.00 a.m. (it was probably later) when the wind died right down and he thought the tropical cyclone was over. The power was off in Trower Road, they had had several windows blown in, and there was glass all over the floor and water on the floor in the hallway. Chris looked outside briefly; he could see stars and some clouds. The air seemed very still, as the wind had died right down. Chris shone his torch out into the yard and noticed that the roof had gone off the kids' bike shed but his boat was still there. His neighbours' windows all looked to be intact. The sense of calm came from being in the eye of the cyclone; twenty minutes later the tempest resumed.

South of Trower Road, at the airport, John Auld was on duty at the Qantas office for international flights. He and some staff were standing in 30 centimetres of water in the terminal building. It was too dark to see outside to check the havoc being wrought upon the light aircraft on the aprons. John was thinking about his wife, Helen, and their baby, Glenn, who were several kilometres away in Alawa. Helen was lying down in the main bedroom when, just after midnight, the glass louvres blew in and showered her and the baby in glass. Helen had some glass stuck in her leg, but she 'was okay' and so was her baby son. She got up, cutting her feet in the process, and found that she was unable to open the bedroom door. The wind was obviously pushing against it and she lacked the strength to prise it open. She was worried about her pet budgie out in the kitchen and kept trying to open the door. During a lull in the wind she succeeded, and raced down the hallway and retrieved the bird.

The best place to shelter, Helen decided, was the toilet, which was separate from the bathroom and opposite the bedroom. She sat on the toilet with the lid down, dressed in her street clothes and holding the baby. The budgie's 'eyes were like saucers'. She had a candle and a transistor radio. The floor was wet and discoloured by the blood from her feet. She could hear the water in the toilet, but as the eye approached the house the water level dropped out of the pedestal. In their totally dark house, alone with her baby, she felt very afraid. The wind was buffeting the house, and she couldn't hear herself think above the noise of the gale, which she described as 'shrieking and howling'. Added to that noise was the constant smashing of flying timber, bricks and corrugated-iron roofing into the house, which made her very frightened. With her baby in her arms, she braced one foot against the toilet door and rode out what she thought was about an hour of very strong gales in the first wind.

The night was black, but huge sheets of lightning blazed and lit the air for a second before Helen was plunged back into inky blackness. Just before the first wind eased, enormously strong gusts slammed into the house and the roof went. Sheets of rain engulfed the ceiling above her, and just as the eye approached, the ceiling collapsed and half of it came down over the toilet door. Water dripped constantly over her from this point on. She now had to sit on the floor and hold the door closed with her back and her legs braced against the toilet pedestal. The plaster ceiling was leaking water, because the corrugated-iron roof was being stripped off the house.

Kevin and Karen Jurek and their three girls aged six, four and six weeks were gathered in one bedroom in their brick low-level house at Jingili. Karen recalled, 'The wind coming in was really loud. It was a noise I'll never forget.' The Jureks could hear the roof ripping off their house amidst a constant roar of wind. Every now and then the house would shudder when it was hit with incredible gusts. The floor was covered in water. Kevin shoved mattresses up against the louvres. Karen was under the bed with the two older girls, while Kevin got into a corner with baby Megan in his arms. Kevin recalled, 'It was freezing cold. Karen was dressed in a nightie, and I had a footy jumper on.' They admitted later that they had picked the worst room to find shelter, as it had louvres all down one side and they were drenched by the rain getting in. Like several low-set buildings around town, the concrete-walled low-set houses generally fared well in the storm, although not all low-set buildings escaped destruction. The Jureks could see nothing from their refuge; it was pitch black, and their constant companion was the roaring wind and hammering rain. They believed their lives were in danger, but to avoid frightening the children when they talked about the storm, they spelled out any words that might frighten the children. One question upper-most in the children's mind during the height of the raging torment was whether Santa was still coming. At one point when the noise was very frightening, one of the girls went into shock and had a fit.

Another low-set house was a concrete-block dwelling occupied by Judith and Gary Watson, who lived in Amsterdam Court in nearby Wagaman. They had made a shelter in their main bedroom when they realised the savage power of the storm. After only twenty minutes of the first wind, the lights went out and their three girls, aged eleven, nine and seven, were immediately petrified. From their shelter they heard the front door blow in, followed immediately by the sound of their fish tank smashing. Once the front door went, the louvres in the lounge room were blown out. This was now too close to where they were, so the entire family, including two cats and a dog, moved into the bathroom where they sat on a foam mattress. Judith had packed an airline bag with a first aid and emergency kit and a change of clothes for the girls.

The bathroom was very small and contained a square, shallow bath. Judith and Gary had to yell at the top of their lungs to hear each other over the noise of debris smashing into the house. They had the windows open on one side to relieve the air pressure inside the house. Gravel from their neighbours' yards was hitting against the window like shrapnel, so Gary stood a mattress up against it. Lightning was flashing brilliantly and constantly through the room, changing it from pitch black to almost daylight, and back again. Then, with an enormous bang and without any warning, the house imploded. All of the remaining windows in the house blew in at once, sounding like an enormous cannon going off. The medicine cabinet above their heads had its door sucked off the hinges and the contents fell on everyone below. The family huddled closer together, protected by the remaining cement walls of the bathroom and very little else.

John and Janice Woodcock realised that they needed to seek the shelter of their hallway when the noise of the storm rose to a deaf-ening roar and debris started smashing into the outer walls of their house in Nakara. At around 11.00 p.m. they started losing louvres and the number of impacts on their house increased. John said he was 'stunned by how strong the wind was'. They knew that the bathroom was recommended as the best place to shelter, but Janice said they decided to stay in the hallway because the bathroom was too small, especially for John's father who was a big man weighing around 150 kilograms. The roof went early in the first wind; soon the plaster gyprock ceilings started falling in and rainwater began pouring through. John tried to lift some of the ceiling off where they were sheltering, but found it to be incredibly heavy. With the roof gone, the damage level rose exponentially and their two young children, aged two and four, were starting to look worried. Windows began breaking and then a falling ceiling light hit Janice. John decided that they would be better off downstairs.

When a gust eased, the family quickly scrambled down their back stairs and went into the concrete block open laundry and storage area. John was illuminating the way for everyone with a 'Big Jim' torch, but he remarked that at times it wasn't needed owing to the 'incredible lightning'. He added, 'Roofing and iron sheeting of the commercial Spandeck type in lengths of about 20 metres was flying everywhere and creating a terrible noise.' He grabbed a mattress out of their car for the children. The pets were also tagging along and a very cold cat and dog crawled in underneath their bodies. John and Janice, John's parents (who had been staying in their caravan which was parked in the yard), the two children and the animals stayed sheltering from the first wind in the open laundry where they were subjected to very cold wind and rain for the next three hours. Janice recalled, 'We could see stuff going past, like fridge doors and roofs, and it was all coming from the north.' Janice admitted she was very frightened for her young family. Even before the roof had gone and the ceiling collapsed on them, the house had been rocking. 'It was very unnerving,' John said. 'But I didn't think it would come down on top of us.'

While they were huddling in the laundry area, John shone his torch out to where his parents' van had been anchored in the backyard. It was gone. His parents were devastated; all of their precious belongings were in the van. The wind changed direction slightly and everyone was now being hit hard by stinging rain, so the group moved inside the storeroom. They hadn't been there long when Janice and John noticed that the leading edge of the storeroom wall was pulsating with the wind gusts and looked as if it might blow in on them. They had no option but to move back to the lee of the outside wall and brave the elements.

At 12 Wanguri Terrace, in Wanguri, Rowan and Marcia Charrington had been discussing the likelihood of a trip to the dump on Christmas Day to retrieve their neighbours' clothes, which they had accidentally thrown out with the rubbish. They needn't have worried; not only would the dump be closed, but Darwin itself would look like a vast waste area by Christmas morning.

In their typical government high-set house, Rowan and Marcia went to bed with a radio, torches and emergency kit next to the bed. But before long the house started to move with the north-westerly gale and water started coming in through the metal and glass louvres. Rainwater was pouring into the passageway and all over the hardwood polished floors. Rowan grabbed towels out of the linen press and started helping Marcia to soak up the water in the hallway. Near midnight, as the wind gusts increased, the house started to move. Rowan said, 'It felt like it was shaking off the pylons.' The house normally shook a little even when someone walked up the stairs; now it was doing that, 'but much, much worse. The house was twisting. There was no crashing or banging, just very strong wind, with lots and lots of rain coming in on one side through all the louvres.'

The Charringtons were now very concerned for their safety. The house was shaking and creaking very loudly as it twisted on its steel pylons. Marcia recalled, 'We moved to the middle bedroom, and opposite was the bathroom/toilet and that was the widest part of house.' They were lying on the bed dressed in shorts and T-shirts, and slip-on shoes and clogs. They were staring at the ceiling when the manhole cover in the bedroom disappeared up into the roof cavity. Rowan remarked, 'We could see the iron roof moving and lifting through the manhole. The hot water tank above us also moved. The wind was now very strong.' Rowan got up off the bed when a lot of noise started down near the lounge and dining area, and he checked the front door to make sure it would hold. Just as he was doing that, there was a huge bang and all the lounge-room louvres blew in at once. A neighbour's roof came smashing into their house with an enormous gust of wind, creating absolute havoc. Rowan's face was impregnated with glass and he was bleeding. He 'retired hurt' back to where Marcia was waiting nervously. Thankfully, it was dark and she couldn't see his face. The sound of the wind, and of things breaking up and smashing, was so loud that talking was difficult; they had to cup their hands over each other's ears to communicate.

'I now felt we were in extreme danger. Survival was now uppermost in our minds,' Marcia explained. They felt the house was going to disintegrate, but owing to the direction of the wind and the debris, they couldn't get out the front door. Everything in the lounge room was shredded, and the door was jammed. They tried to remove the metal louvres from their frames so that they could get out on to the stairs to escape. The wind was still strong on the lee side of the house but was now also coming up from underneath. They decided to go into their downstairs storeroom, for which Marcia had the key. They had made an exit from which to climb out on to the stairs. Rowan went first. A heavy wall unit near to where they were climbing out was rocking in the wind. Rowan stood on the stairs waiting while Marcia started climbing out with her kitten in a carry-bag along with their emergency kit. Just as she came out on to stairs, the wall unit guillotined across the cavity. Had it hit her, it would have broken her back or possibly crushed her to death.

The first wind was still raging. It was so strong, the pair couldn't walk down the stairs, so they slid down on their backsides and then managed to crawl and clamber their way into the laundry. This area was shaped like an 'L' and built of concrete blocks. It was solid. Inside was a very solid workbench that Rowan had built. He grabbed a sheet of plastic to keep off the rain which was pelting them through the gap between the top of the wall and the floor joists. Rowan commented, 'Down inside the laundry area at ground level, the noise was incredible. As we were climbing out, I noticed that the end of our kitchen and the adjoining verandah had gone.' What he didn't know was that the entire end wall and master bedroom had also gone. Their decision to move had probably saved them from death or serious injury.

Rowan recalled, 'There was heaps of lightning. In fact, I could see *too* much!' They stayed in the storeroom in the dark under the wooden workbench which they were confident would withstand even the block wall coming down on them. Marcia said that she 'tried to imagine it wasn't as bad as it sounded'. They spent the next three hours feeling very nervous; they were wet and freezing cold. They couldn't see much around them or outside of their shelter. Rowan commented that their 'greatest concern was our fear of the unknown. Just how much more damage was going to occur?' For the moment, though, they felt that they wouldn't be crushed. Their car, parked next to the laundry under the house, was providing some protection, and the concrete block walls seemed to be holding. In the myriad flashes of lightning, they could see the floor above them moving. Rowan noticed that occasionally he could see light *between* the blocks in the wall. They clung to each other under the bench— cold, cramped and praying for the wind to die down.

About 45 kilometres out of town and to the east of The Track is the remote village of Humpty Doo. In 1974 it contained only a dozen or so houses. These hardy souls lived out in the bush on acreage with no light, no power and no water; they used rainwater and septic tanks. The local pub at Humpty Doo was also the Civil Defence centre and general store of sorts. Jenny Lonergan's husband John had made 'a right mess of himself' at his Christmas Eve work party and gone to bed. His favourite garlic meatballs were eaten by their pet dog, Goldie. The wind hit the tiny settlement much later than in town, having taken quite a few hours to travel down to the southeast. The Lonergans had a house on 10 acres in Cypress Road. During the gale, a huge 20-metre-tall blackbutt tree in their front yard fell on to their bedroom. Jenny and John crawled out from under the wreckage and found the door jammed. John bashed a hole in the door and they climbed through and then got out of their wrecked house. They had already had a close shave when the ceiling fan in the bedroom had nearly impaled them on the bed.

They got dressed quickly and jumped into Jenny's lightweight Suzuki four-wheel-drive car, along with Goldie, and headed off to find shelter at the pub. The owner, Skews, wouldn't let them in, and so they sheltered between a couple of road trains that were staying there before moving off to Maningrida the next morning. The truckies helped them to tie down the car and they then tried to settle down for the night. Because the Suzuki was likely to be blown over, John and Jenny were wearing their motorcycle helmets. Goldie's dinner of garlic meatballs now came back to haunt them. Jenny wasn't sure which wind was worse, the tropical cyclone or 'bloody Goldie farting all night'. The wind was ripping trees out of the soft, waterlogged ground, and Jenny recalled seeing a tree moving upright a metre and a half above the ground with its roots dangling below it. 'It was dancing along the road.' She added, 'The noise was deafening, like an evil scream.' Jenny admitted that she was scared when the wind hit Humpty Doo. One sight she said she will never forget was 'the chook shed at Skews' pub. A whole cage on four legs, 5 x 1 metres, bounced across the road still with all the chooks on board. Remarkably, they all survived.'

At around 4.00 a.m., Jenny saw what she believed was the eye of the cyclone. 'It was a big orange glow in the sky, a sort of yellow light that was moving around—an evil light.' The eye stayed overhead for about twenty minutes and then the wind returned.

The first wind was slowly easing as the centre of the tropical cyclone approached. The first wind had done a fair amount of damage, especially to those properties that were exposed or in the northern suburbs where there was nothing to stop or deflect the power of the wind. The ground was littered with rubble, debris from wrecked houses, and every manner of rubbish, from broken windows and louvres ripped out of wall mounts to pieces of fibro that had blown out of walls. Anything not tied down inside a room, from lounge chairs to cushions, lampshades, pictures on the wall, ornaments, records, pot plants and even floor rugs, was blown or sucked out through holes in walls or broken windows. There was an enormous amount of roofing iron and timber that had been ripped off roof trusses or from other parts of buildings. Fences were pulled out of the ground when lengths of roofing iron were forced up against them, acting as a lever for the wind. Children's bicycles, rubbish bins, pot plants, garden tools—anything that could be blown around, had been. All of the debris lying around now became a factor in what happened when the eye of *Tracy* passed and the wind changed direction.

8 EYE OF THE STORM

I T WAS NOW well after 3.00 a.m. on Christmas morning. In some areas of Darwin the eye didn't pass directly overhead but off to a flank, and as it did so the wind changed direction. The eye of a tropical cyclone is like a huge chimney, and inside that hollow column there is very little wind or cloud, and no rain. Consequently, there is also very little noise. It is an area of extremely low pressure. In *Tracy*'s case this has to be estimated; the last reading before the barograph was damaged was 955 millibars. For those who had the eye pass directly over them, it was an unsettling, and in some cases a dangerous, time. As it moved inland, the eye slowly started to fill with rain and moisture, as it was unable to draw any more heat from the earth's surface, which was much cooler than the sea surface. Post-cyclone images indicate that as the tropical cyclone passed the western edge of the airport, which is 8 kilometres from the coastline, the diameter of the eye shrank from 13 kilometres to around 9 kilometres.

The eye missed the Darwin CBD and some of the inner suburbs. Eddie Quong, in Shepherd Street, recalled that it was 'more of a lull, not really a proper eye'. At Larrakeyah Army Barracks, Peter Kerntke looked outside:

> It was quite eerie. It was also a bit upsetting, because the soldiers knew that the wind was coming back. I could see quite a lot—probably because of the lightning—and I saw bare palm trees around the base. The eerie quiet was actually scary. It stayed for about what seemed like fifteen minutes or so.

Over at Stuart Park, Greg Huddy recalled, there was 'peace and quiet'. He knew the second wind was coming, but he thought they had had the worst of it. From the bedroom where they were sheltering, he couldn't see much outside; it was 'so dark and black'. He thought the eye was overhead for twenty minutes. His wife Barbara described the eye as 'very deathly quiet'. Greg added, 'It was a perfect eye—absolutely calm, no rain and no wind.' (Given that they had been lashed with winds of up to 150 kilometres per hour, it is understandable that the slight breeze typically found in the eye of a cyclone passed for dead calm.) Cherry Perron, a kilometre away, recalled seeing a strange green light at the time the wind changed direction.

At Ludmilla, Alan Grove's three young children slept through the first wind. Alan could see outside his house during the flashes of lightning, which he said gave off a 'brilliant light'. He recalled that where they were, in the protection of a small rise, there wasn't a great deal of debris lying around after the first wind. He ventured outside:

> It was eerie being inside the eye. It felt deathly still, with no air, and I could feel the low pressure. I met my neighbours and we chatted. The house across the road was badly damaged, but the owners were away. One house up the street had also lost a room off the end of the building. It was strange and unsettling, because we all knew it [the cyclone] was coming back.

Qantas traffic officer John Auld, out at the airport, thought the eye was overhead at around 3.30 a.m. He didn't move out of the office where he and the staff were sheltering. He thought the eye loitered for up to thirty minutes. He recollected, 'I thought about getting in my car and trying to drive back to Alawa, but I knew from what the mets had said on the telephone that the second wind was coming.' His wife Helen, who was home alone in Bald Circuit in Alawa, was frightened for her life. She considered sheltering in the car, but decided not to take the chance; she wasn't sure how long the calm would last. The eye could remain over her house for twenty minutes, which was the average period of calm between the first and second winds, or it could move on after just ten minutes, bringing more devastating winds in its wake. Helen decided she didn't want to be half-way down an outside staircase covered in sharp, jagged corrugated iron and other debris when the wind returned.

Many people used the word 'eerie' to describe the atmosphere when the tropical cyclone's eye passed over. Dianne Ferguson, in a married quarter on the RAAF base, recalled, 'It was completely still. I could hear voices. I heard someone yelling for help and a woman screaming. My husband Bob wanted to go and help, but I wanted him to stay because I knew the wind would come back.' Bob took a torch and went outside and looked around. When he returned, Dianne remembered, he sat down and was very quiet, saying only, 'There's glass all down the hallway.' Dianne wanted to use the toilet next to the bathroom where they were sheltering, but there were no walls left on the toilet—just the pedestal. She thought it was 'like an autumn morning. The calm before the storm.'

Just up the road, in another married quarter, Josephine Foreman recalled:

 The air was so hot and still. It went from a roar to absolutely dead quiet. We could hear neighbours screaming. I couldn't stand up because my legs were cramped from being stuffed into the shower recess area. There was a strange light, and the lightning in the background reminded me of St Elmo's Fire.

Josephine forced her husband to go and help their neighbours. Others also came along to help people trapped in the wreckage of their houses. During the thirty minutes that Josephine thought the eye lingered, the neighbours got in their car and drove to the cyclone shelter at the RAAF base. True to form, Josephine's husband took the opportunity the calm provided to get some beer from their fridge.

In the coastal suburb of Nightcliff, Chris Collins and his wife Trisha were in their bathroom. When the eye arrived, the noise died down very quickly. But they couldn't hear much around them. Chris placed some tarpaulins over their gear in the lounge and bedrooms, and he and Trisha had a drink and some Christmas cake.

They suspected that they were in the eye, which they estimated lasted between ten and twenty minutes. 'It was enough time to look around by torchlight,' Chris said. Sheets of lightning were still illuminating the sky. They were very cold, and it seemed to be getting colder, which Chris reckoned 'was a function of the lower pressure'.

Judy and Tony Pickering and the dozen or so partygoers in Clancy Street, at Fannie Bay, recalled the eye distinctly.

> The wind stopped, the blokes went outside, and everyone was really worried about the next wind. One of the men [Ian Britten-Jones] was a pilot, and he said he knew it was the eye. I saw that the upstairs of the house was almost demolished, but the yard wasn't totally strewn with trees or timber and metal. However, the cyclone wire fence at the end of the yard was packed with stuff. A large rain tree had fallen across the in-ground pool and covered it.

There was a group discussion, and after Ian Britten-Jones warned about the coming second wind, the men retreated back into the large rumpus room to await the storm's return. Peter Coombe had returned from the hospital with the people from his own house in Ludmilla. They used a suitcase as a bassinet for the injured woman's very young baby. Tony Pickering recalled that the husband of the woman who had been injured 'lost the plot' and was probably in shock.

Milton Drew was home alone in a very empty house in Fannie Bay when the eye came over. He said, 'Everything was perfectly, deathly quiet and still. I'm not a religious person, but I was almost blasphemous. I said out loud, "If that's the best you can do, I'll be okay."' Around the corner, Elspeth Harvey and her family were sheltering in their car parked under the house where they were partially protected by a large trellis of flowering vine. During the eye she ventured out and found the missing budgies in their cage well down the backyard. Elspeth recalled:

> It was dead calm, still raining lightly and there was quite a bit of lightning. I looked around. Before then, I'd thought it was only our house being damaged. I saw that our lounge room was smashed open. There wasn't

much rubbish in our backyard. My son Gordon's boat was still there upside down—undergoing repair. I had a 'Big Jim' torch, and shining it around quickly I saw that we weren't alone; other houses had been damaged too. Then Gordon came around from a flat he was living in at the end of the street. He saw that we were okay and then ran back to his flat. I found the pre-school's budgies—but still no cocky.

On returning to her car, Elspeth covered the budgie cage with a towel, 'because the cat was eyeing off the birds!' They settled back in the car and waited.

In Croker Street, in the northern suburb of Nakara, John Woodcock was feeling the intense cold and, like many people, felt a great need to urinate during the storm. He got up to relieve himself when the eye—or what he and wife Janice think was the eye—passed overhead, but he found it extremely difficult to move around because of the amount of debris on the ground around their highset house.

Judith Watson, who lived close by in Amsterdam Court in Wagaman, thought that the wind just 'lulled' a bit. She and her husband were unable to open their bathroom door because the house had imploded and enormous amounts of rubble and debris kept it jammed shut.

The Charringtons, in Wanguri, remembered having an eye of sorts pass overhead. Rowan said it was 'quieter, with not as much wind'. He added, 'There was no tearing of iron or ripping of timber and crashing.' The couple thought that the wind 'howled a little, but the gusting had stopped'. Rowan wanted the two of them to move from under the workbench in the storeroom, where they were very cold, and into the car, which he thought would provide more protection. Marcia wasn't so sure; she had heard warnings on the radio telling the public *not* to shelter in their cars, especially under houses. Rowan decided to have a look around, though, despite Marcia's protests. He was sure that they were in the eye of the storm. He recalled the scene:

 During a flash of lightning I saw our close friends' house at the back of ours. It was totally demolished; all that was left was just piers and three RSJ [rolled steel joist] bearers—no floorboards. They were the Buchanans. I thought that they were probably dead. There were big chunks of floorboard sections in our backyard. I could see that their downstairs blockwork was gone. I quickly ran up our back stairs and saw that the entire end of our house was gone, and I returned to Marcia.

After seeing the damage to their friends' house, which had caved in on the Buchanans' car, Rowan decided that they would be better off where they were. They would just have to 'sit it out and hope like hell we wouldn't get killed'. Marcia thought there was 'a good chance we were going to die'. Rowan said he was greatly shocked when he saw that the Buchanans' house was gone. He didn't tell Marcia, but she said that when he came back and got in under the bench, 'his face was like stone and he didn't speak for thirty minutes'.

The eye was a salvation for quite a lot of people, as it gave them an opportunity to relocate to safer shelter. Others mistimed their window of opportunity and came to grief when they were caught out in the open by the second wind.

9 ROAR OF THE DEVIL

THE FIRST WIND had created a significant amount of damage everywhere it hit. It was very strong and had probably blown consistently at around 120 kilometres per hour, with gusts to 150 kilometres per hour, as the mets had predicted. But the winds that came from the south-east in 'the second blow' seemed to most of the eyewitnesses to be stronger and more powerful. Graeme Clarke said that one factor that made it *seem* worse was that the second wind, which many described as a 'roar', came out of the relative silence of the eye. In fact, the second wind was indeed more powerful, with the highest wind recordings occurring at this time and gusts of around 280 kilometres per hour likely to have been achieved.

The wreckage and rubble from the first wind was now lying on the ground, and when the second wind hit with a sudden onslaught it carried all of that debris with it. Not only was the wind now stronger, but it contained missiles that would wreak havoc on people and buildings alike. The first wind had built up over an hour or more, but the second wind arrived with a suddenness that shocked and terrified everyone in its path. It was now a couple of hours before first light, and for many people in Darwin those hours would seem interminable. For some, they would be their last.

Diane Kerntke described the duration of the second wind as 'the longest night of my life'. All of the windows in the solid cyclone shelter of the Officers' Mess at Larrakeyah Army Barracks blew in, and everyone in the room was very cold. The horizontal rain 'soaked everyone through to the bone', Diane said. Her husband Peter was on duty in the army's operations room, which was still able to function, as the emergency power generator had kicked in. When

Peter went to check the flooded generator room, he found that 'the room was totally dry—the water had all been sucked out. The wind had created a vacuum and several hundred gallons were gone.'

In Shepherd Street in the CBD, Eddie Quong and his family were being pushed around by the storm. They moved under the kitchen table, which they had dragged into the hallway. Eddie was using a piece of timber to hold a section of wall off his body, as he was unable to get under the table. When the wind hit 'with a thump', it removed the master bedroom, part of the roof, the kitchen and the bathroom, which were directly in its path. Greta recalled, 'I thought I was going to die. The noise was so loud, we couldn't talk.' Everyone was filthy from the dust and water showering down on them from the ceiling. The second wind lasted until around first light, which was probably close to 6.30 a.m. It was nerve-wracking and exhausting, and Greta fell asleep holding on tightly to her two daughters.

The people sheltering in Kelly Morrissey's rumpus room in Clancy Street, Fannie Bay, went quiet after the eye passed. Tony Pickering looked around the room and saw some people quietly praying as the second wind smashed ferociously into the remains of the house. The men strained against the mattresses that were holding the glass doors that led out to the pool area. Luckily, no serious flying debris smashed into the doors. The rain tree was providing a screen of sorts and filtering out the 'nasty stuff'. Above them they could hear large objects—some thought they were probably refrigerators—bouncing off the floorboards. Bricks and heavier missiles pounded against the brick walls of the rumpus room for the next hour and a half. Tony Pickering said, 'It was a tense time. We were just hanging in and praying that the mattresses would hold, and they did.' Everyone was exhausted. The Pickerings recalled that the second wind 'was the scariest part—the noise. The worst time was after the eye.' The huge rain tree that was over the pool was thrown back past its original position. Hanging over the floorboards at a 45-degree angle was half of Kelly Morrissey's house. Judy Pickering thought they would be the next to go.

Elspeth Harvey was sitting with her family and menagerie in her brand-new car just around the corner. She felt that at any time the small car might be rolled over. The trellis that had protected the car was now blown back up by the second wind. Elspeth recalled a nasty moment during the raging storm:

> I wanted to pee really badly and I wouldn't do it in the car on my new seat covers, as my husband Alex suggested. So I got out and had the fastest pee in the world, hanging on to the car door. But while I was hanging there, the bloody cat escaped and stayed out. The second wind was worse; it had a frightening noise, a roar. I was scared, but I didn't panic, I guess because I'm a fatalist.

The Harveys waited in Elspeth's car until first light, but it was facing into the laundry wall and they couldn't see much. All the vegetation was being ripped off the bushes and trees, and the entire car was splattered with leaves and bits and pieces of rubbish.

At Ludmilla, near the airport, Alan Grove was wondering where to go next when the wind came back. He recalled:

> It hit within a minute, and it was a sound much deeper than the first wind—like a heavy roar. Instantaneously, we knew something catastrophic was coming. We put the kids in the lounge room between some mattresses, but that very quickly got too wet and so we moved to the bathroom, and then moved again when a tree came through the side of the house.

Luckily for the Groves, they were in the lee of a small rise and the worst of the wind went over the top of their house. They had massive debris damage to the walls, but the house survived. Alan Grove recollected a funny incident during the storm; as he said, 'There weren't too many of those.' As his glass louvres started breaking, he thought angrily to himself, 'Where am I going to get a glass louvre tomorrow?' Alan thought the second wind was 'at least 50 per cent stronger and noisier than the first wind'. The high-set house was vibrating, and rain was being forced through the remaining louvres. After the tree smashed into the bathroom, the family could see outside and could easily be hit by flying objects, so they moved again into the passageway at the rear of the house. They were now furthest from the wind and, as Alan hoped, 'in the strongest area'. He recalled:

> My kids were traumatised. The roof stayed on the house, and we hung on in the hallway and didn't move around too much after that last move. The wind lasted and was really bad until at least 4.00 a.m., and then we noticed a lessening in wind strength. It was a bad time. I'm glad we couldn't see outside. We were among the fortunate few who survived with a house sort of intact, but I felt that we could all be killed.

On the RAAF base, Dianne Ferguson was frightened for her life when the second wind came. While the family huddled together in the bathroom, she started praying. Her son Michael sat quietly; her other son, Kerry, 'whinged a bit'. Dianne recollected that the second wind 'came as a blast, not like the first wind which had built up. It was quite sudden—like a car crash.' The walls in the bathroom began to flex and Dianne and Bob became very worried. Bob grabbed a blanket and put it over their heads. Dianne added, 'I prayed harder and sang louder.' The wind was now so loud they couldn't hear the rain. The bathroom windows were rattling and some smashed, but the blanket protected them. The bathroom walls stayed intact, but Dianne noticed as it became light that the water in the bath was filthy from airborne dirt. She thought that the worst part, and what frightened her the most, was 'hearing the explosions, the noise of the wind and the impacts' on their house.

Just around the corner, Josephine Foreman also heard the second wind coming. She recalled:

> There was no power, it was dark, and our kitchen and back wall was gone. There was nothing behind the cupboard doors! I climbed back into the shower recess and put my kids on my lap again. My husband barricaded up the louvres in the shower. It was an incredible noise, as stuff like tin was rolling down the road. The kids were very scared. There was a big bang, and a bearer came into the shower and impaled itself straight through the bathroom where we were sheltering and into the other wall.

Josephine now thought they would be impaled to death. The married quarter was twisting and shifting. She thought that the wind would now get under the house, weaken it and they would fall through the floor. The empty house behind their quarter imploded, and all the wreckage smashed into their house. The wind was now coming straight on to the bathroom wall. Josephine knew that she had been cut by glass but was unable to do too much about it for the moment. She felt sure they wouldn't survive. She said later, 'I prayed and said to God, "Just make it quick. Don't let us suffer."'

At Nightcliff, young Nat Guinane recalled that the wind changed direction and was stronger. She said, 'It hit quickly. The noise was worse. It ripped one way and ripped it back again and brought more debris.' The concrete stairs on the two-storey block of flats were ripped off the wall. The roof went off the top flats, and most of the concrete block walls disintegrated to varying degrees.

At Rapid Creek, Chris Kingston-Lee and his family hadn't had too hard a time of it through the first wind and thought they would be okay for the rest of the night. That soon changed. Chris was lying down on his bed, feeling very tired.

> " I was just about to go to sleep and I heard this semi-trailer coming down the road. Then forty more, and then a hundred more, and I wondered what the hell it was. It sounded like they were coming at us flat out, revving their guts out. Suddenly, the wind slammed into the house. I grabbed my son and the blankets and we went into the bathroom. My wife and I couldn't talk to each other, and I thought, 'Holy shit!'

The roof then started peeling off, taking most of the ceiling with it. Chris said he 'distinctly heard the bloke next door yell out'. Roslyn was under the bathroom sink, and Chris covered her and their son, 'TJ', with blankets. The dog was behind Roslyn and huddled in under the sink with a look on its face that said, as Chris described it, 'You can bite your arse. I'm not moving!' And the dog stayed there. Chris was still only in his underpants and sitting in the small bath. 'Fibro sheeting fell through the open ceiling and into the bath, and so I covered myself with that,' he said. 'The noise was just incredible; it was whistling and screaming. The rain was hitting us like we had the shower full on in the room.'

By morning, the family and the dog were in water 15 centimetres deep. At one stage, Chris said, he looked up at the sky and 'could see stuff flying by. Another roof landed on our house, so we had no rain for five minutes, and then it left again.' Debris was now blowing in and out of the open-roofed bathroom. Everything was blown, or more likely sucked, out of the bathroom vanity, including the door. Roslyn (whom Chris described as normally being demure) pulled the blankets aside, looked up and asked Chris, who was shivering in the bath, 'How long is this going to fucking last?' Chris replied that he thought about another twenty minutes. About an hour later, Roslyn came out from under the blankets again. Chris saw her frightened face lit by the lightning flashes as she screamed, 'You're a fucking liar! . . . I hate you, God!' For Chris, the incredible raging wind was the worst part, especially when he could hear the house breaking up. Eventually, he got out of the bath and huddled up close to Roslyn on the bathroom floor. During a small lull she said in his ear, 'If I don't make it, remember I love you.' They both thought they might be killed.

Karen Jurek, who was sheltering in their low-set concrete house in Jingili, was having exactly the same thoughts. When she heard the roar of the second wind, she said, 'I thought that this was it—we would die.' Sheets of iron were now tearing off their roof. In fact, the entire roof was lifting off the pre-cast concrete walls, including some of the trusses. They also spent the night sitting in water, but, as Kevin Jurek noticed later, all nine houses that he knew about of similar construction basically survived the storm. When a gust of wind came, they could hear it accelerating, and there was one every couple of minutes or so. Each time, Kevin wondered 'if this was the one that was going to get us'. He wanted to get into a cupboard, thinking it would offer more protection. 'We learned to pray,' he said.

In Alawa, Helen Auld could also hear the wind gusts coming. She was terrified and began to pray. Eight weeks pregnant, she was worried about having a miscarriage.

In nearby Nakara, John Woodcock and his family and parents were forced to shelter out in the elements, using the lee of a concrete block wall as their only cover. They were being pelted with rain and mud. John explained that they had to shelter where they were because their cars were directly out in the wind and had lost all their windows. They could see out from where they were huddling when the lightning flashed. John recalled:

> There were huge gusts that lasted for three or four minutes. Objects like refrigerator doors flew past in these gusts, which were much stronger than the screaming wind. All the trees were flattened; nothing over 2 metres was left standing in the backyard. Our neighbour's Landrover was upside down under his house between the pylons.

Incredibly during this time, they could hear above the noise of the roaring wind a woman screaming. One of their neighbours, they learned later, had been wounded in the knee by shards of glass. Only one upstairs room of their house survived; the rest was torn to shreds and holed. Almost the entire roof of the Woodcocks' house was peeled back, and all the walls were gone except for the third bedroom. Everything that wasn't nailed down was sucked out of the house. Their cars were pushed around the front yard. Janice said she would never forget 'the noise of roofing iron scratching against concrete'. For a long time, the metal roofing that was wrapping around power poles and trees was caught very close to where they were sheltering and only half a metre from their feet. One of John's strongest memories was of 'a passing fridge stuck on our back stairs'. The two caravans parked in the backyard were tumbled 40 metres across the lawn. Despite all this horrific mayhem, at one stage one of the kids asked if Santa would still come.

Judith Watson, in Wagaman, said she was 'glad' she couldn't see outside during the storm, because it was scary enough as it was. When they tried to get out the next morning, it took them thirty minutes to get free of the rubble.

Rowan and Marcia Charrington, in Wanguri, thought that the 'quiet time' lasted for about ten minutes before they could hear the crashing starting again. Rowan said it sounded like 'a squadron of tanks starting to drive over all the wreckage'. It hit the concrete block storeroom where they were hiding under a workbench 'with a whump', Rowan said. 'We heard a big thump and something hit the block walls and it all came down.' They thought they would die, he said. The wait in the dark seemed interminable. Marcia said that she lost all concept of time, but she had 'faith in Rowan's bench'. She added, 'I made him promise he wouldn't go outside again.' The young married couple rode out the second wind under the bench, with no walls to protect them. Marcia said, 'I wanted it just to stop. I was thinking, "Where is it all coming from?"' They were now totally exposed except for the cover above their heads. They started to feel very cold and numb, and were cramping up from poor circulation. Marcia admitted that she wanted to pee constantly. All they wanted, they said, was for the day to break so that they could see again.

In Wanguri, Bill Gough recalled that the noise was very loud. He compared it to a jet engine hovering over the house. Looking out into his backyard through the toilet window just before the second wind hit, he saw a 'greenish light' and recalled, 'When you see trees half a metre across being pulled out of the ground like weeds, it's cause for concern.' The roof of the house went, and most of the walls disappeared. But they survived, although they were very wet and extremely scared.

Rick Conlon, who lived at Howard Springs, 30 kilometres out of town, said afterwards of the wind, 'It burnt the country for 15 kilometres miles each side of Darwin. It stripped all the leaves off everything. The trees looked like they were dancing. They were ripped out by the roots and moved along the ground.' Rick recalled that they didn't get the eye, but at 3.00 a.m. *Tracy* seemed to be heading due east and had started to wane. He recalled that the wind changed direction and then came from the west, so it appeared that he was north, and on the edge, of the eye. In the backyard of his house, Rick had a 20,000 litre water tank. 'It was pushed off its stand and started rolling along and it took out the kitchen.' At one stage when he looked outside, he said, 'I thought it was the sun, but it was the sky lit by lightning.'

The family had sheltered under the dining-room table and dragged mattresses in after them. Rick recalled, 'The dog was under my feet with her nose almost up my bum.' Then, in an enormous gust of wind, his 18 x 17-metre roof went in one complete piece. 'The scariest part was once the roof went,' said Rick. 'We were very much aware of our vulnerability from that time on. I guess, also, there was the fear of the unknown.' The floor, walls, windows and doors of their house stayed intact, as did the ceiling, although it was full of water and the ceiling fans were almost touching the floor in some rooms. The family moved into the laundry on the eastern side of the house. Everyone was now very cold and wet, and Rick busied himself by trying to roll a cigarette. The Conlons' six-year-old son Gregory wanted to know what was going to happen to Santa Claus. Ten-year-old Jennifer was traumatised by the events of the night. Rick and his wife Cheryl both felt very frightened and apprehensive. 'I wasn't panicky,' said Rick. 'But I was still scared.' The family prayed.

10 RIDERS OF THE STORM

MOST PEOPLE WHO survived Tropical Cyclone *Tracy* felt, at some stage during the storm, that they, their families or their neighbours were in peril. Many thought there was a good chance that they wouldn't survive until dawn. Sheer determination and courage enabled some to struggle through; others were seemingly touched by the hand of fate and survived despite *Tracy*'s best efforts to take their lives. In the midst of their terrible ordeal, couples held hands and told of their love for each other; others relied on eye contact, the roar of the wind making it impossible to be heard. Countless couples have told how they prayed for the first time since they were children kneeling by their bed. Many people cheated injury or death a number of times, moving from one refuge to another as their houses were destroyed around them. It was extremely difficult to move around under the conditions, especially with young children and family pets. It was difficult to see anything; 'It was as black as the inside of a dog's guts,' was how one person described it. The wind made movement above a crawl on all fours extremely hazardous; and anyone moving away from cover risked being impaled, struck, clobbered or smashed into a wall or window. The air was filled with debris, which threatened life and limb at every turn. The wet and slippery surfaces made any movement chancy. The rain was horizontal and blinded anyone trying to look into the wind.

Barbara and Greg Huddy, in Stuart Park, took refuge with their three children, all aged under three, in the upstairs part of their high-set house not long after the eye passed over. Barbara recalled Greg saying it was going to be okay. 'He lied! There was an incredible roar that kept coming, and something—probably someone's house—sheared off the end of our house and the roof then lifted off.' They had initially sought refuge in a middle bedroom, but when that was put under threat they went into the bathroom. Because of the design of the house, the bathroom wasn't viable, so they moved into another bedroom, only to relocate again in the lounge room when the bedroom was struck by debris. At this point, Greg wanted to go downstairs and shelter in the concrete-block storeroom, but Barbara insisted they stay upstairs. Greg admitted later that it was a wise move: the next morning, in daylight, they saw that the steps they had intended to move down in the dark had gone and the storeroom was destroyed. The Huddys then stayed in the corner of the dining room with their backs to the door. Greg pulled the dining-room table over their heads, and Barbara padded their position with cushions. From where they were sheltering, they couldn't see the maelstrom around them. The end bedroom, including the walls and roof of the house, was sheared off by a neighbour's toilet—not just the pedestal, but the entire toilet enclave, including the walls. Greg recalled, 'We burst into prayer—being of good Catholic upbringing. I think we went through every prayer I'd ever learned. But one stupid thing I did was throw a curtain rod to try and open a louvre to relieve the air pressure; we'd already lost 600 louvres!'

At this stage the wind was screaming up off Frances Bay and the family were getting drenched. The Huddys admitted that they were scared for their lives. Barbara recalled:

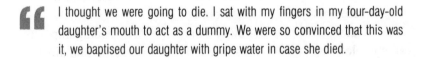

I thought we were going to die. I sat with my fingers in my four-day-old daughter's mouth to act as a dummy. We were so convinced that this was it, we baptised our daughter with gripe water in case she died.

Young Nicole kept saying she wanted to go to the toilet. Greg said, 'I told her just to do it where she was, because we were sitting in 15 centimetres of water anyway.' Barbara had a bad cut on her foot which was bleeding. The noise now was absolutely terrifying. 'It was more fearful—that great fear of the unknown,' Barbara said. 'We were trying to persuade ourselves that it was going to get better.' The Huddys prayed for daylight, believing that things would be all right in the morning. Greg said, 'We couldn't move for fear of being killed. We felt very alone.'

The house was moving an alarming degree off its steel pylons. By dawn the roof was still 'attached', but it had been peeled off and was hanging over the side of the remains of their shattered house. Daylight came and they heard voices, but as Greg recalled, 'We weren't game to stand up until about 6.30 a.m. The wind had started to ease but was still a little dangerous.'

Not far away in Gothenburg Crescent, in Stuart Park, Marshall and Cherry Perron had decided to shelter in a middle bedroom of their high-set house. The position of the house on a slope above Frances Bay, and the wind direction, had made the decision for them, but this time luck was definitely on their side. They had missed the eye, and the wind stayed at peak strength and changed direction. Now it was coming from the south-east and the wind in the storm was at its strongest. Marshall recalled that when the wind increased in intensity and the gusts became incredibly strong, 'the window behind us started creaking and I knew that this was bad! Now it was serious.' They had decided not to shelter in the bathroom, because it was more exposed to the force of the wind and did, in fact, disintegrate.

'The hair on the back of my neck started to stand on end,' Marshall said, when the glass sliding window panels behind where they were sheltering 'started bending' and the window started 'creaking inwards'. They couldn't go out through the door into the hallway to escape this new danger because a whirlpool of glass and crockery was whipping around in a vortex created by the lounge-room and kitchen roof being peeled off. They would also have been out in the elements to an extent, as a lot of the house was being smashed off the floor joists. The incoming wind was pushing at great force into that space and, as Marshall commented, 'The thought of being struck with shards of glass wasn't a good one.' They made a quick decision and got into the built-in wardrobe. They quickly threw out the hanging clothes, and Marshall stood stooped with his legs apart while Cherry sat down with the cat and the transistor radio, which wasn't working. The problem they now faced was how to hold the doors closed when there were no handles on the inside. Necessity has always been the mother of invention, so Marshall grabbed a wire coat hanger, unwound it, and then wound it around the door handles on the outside of the doors. Straining as hard as he could, he then pulled on it from inside the wardrobe, leaving a small crack where he could just see through into the bedroom. Seconds after he had pulled the wardrobe doors to, the window shattered. Marshall continued, 'There was an explosion, like a huge shotgun going off, and once again we could hear the swirling glass going around and around inside the bedroom.' All of the clothes they had just thrown out of the wardrobe, along with the walls and doors, were shredded by shards of plate glass. In the morning, they found that the doors, floors and walls were gouged out where the swirling glass had chopped into them. One minute of indecision would have cost them their lives. They would spend the next five hours inside the wardrobe.

Holding the door handles with a piece of coat hanger wire became a problem. Marshall explained, 'The handles were only small cast alloy, and I knew that if I put any significant pressure on them they would snap off. If they snapped off, I would have two flapping doors I would be trying to hang on to with my hands out there amongst all the glass.' The suction on the doors was terrific, with the wind taking purchase on any surface. As the wind gusted and tried to pull the doors open, Marshall would play out the coat hanger a short distance to release some of the pressure from the door handles while still holding the doors closed. He would then draw the doors back in again as the gusts decreased in strength. He described it as being 'like playing a barramundi on a rod'. Even inside the wardrobe, Marshall said, the noise was like 'a high, deafening roar—like sitting alongside a train that's one foot from your ear'. He added:

 Every few minutes, it was as if a giant hand was shaking the wardrobe and what was left of the house most violently—like an earthquake. And these were terrifying periods. The timbers on the house were being ripped apart, and I thought that any second we would be airborne with our cupboard around us . . . It was quite an awful thought, because we were facing potential death.

This battle with the wind lasted from around 1.00 a.m. until half an hour before dawn. Marshall admitted that it never crossed his mind that anyone else's house was being as badly damaged as theirs. Many people were under the same misconception, believing that only their own house had been unable to withstand the unbelievable strength of the wind that followed the eye. Cherry said that she was cursing the builder all night! The wardrobe refuge was right next to the solid end wall of the house, with the back of the built-in robe being the end wall itself. Inside the wardrobe were all their Christmas presents, one of which was an eternity ring. Cherry had insisted, before she would get into the wardrobe, that Marshall present her with the ring.

Sitting in Cherry's lap was their devoted cat, who was very nervous. Cherry recalled, 'She became quite neurotic afterwards and was never the same again.' Marshall agreed, 'She went quite bonkers.' Meanwhile, Marshall was continually playing the door handles with the coat hanger in a deadly survival game. He remembered 'feeling hot and not too wet from the rain', though the ceiling had been leaking. Most people caught in the storm spent the night freezing, and some children even caught pneumonia. Straddled above Cherry, Marshall was sweating, tense, concentrating hard on his task and had a 'high level of anxiety'. Cherry was also uncomfortable, lying in the bottom of the wardrobe and squashed up with the cat. She said she 'felt afraid of being picked up and swept away'.

The wardrobe was constructed of chipboard and not very strong. But there was nothing more they could do; they had run out of options. If the eye had come, they would have got into one of their cars, risky as that was. The torment raged all night. Marshall said of their time spent in the wardrobe, 'It was a long time to feel that you are going to die.'

Every now and then during this 'most terrifying period', lightning would illuminate outside 'as if it was day'. The Perrons were feeling extremely frightened, as the house was being buffeted and shaken very violently. They reassured each other as best they could. Cherry remembered telling herself not to panic, because 'there wasn't a damn thing I could do about it'. The cat was all claws, she said. Marshall summed up the time they spent fighting the storm from within a wardrobe that could so easily have been destroyed: 'I should have kept that coat hanger—it saved our lives.'

Until the second wind hit, Julie and Heikki Tammo in Larrakeyah had fared reasonably well, though Julie was upset that the roof was leaking and rain was being pushed through all the louvres. Heikki was out of bed, and as they were moving around the house the wind suddenly changed direction. The eye hadn't passed over their house, and now the wind was coming from a different direction. It wasn't fully 180 degrees, but more like 90. Water now started flooding into the house, and a couple of sheets of iron suddenly peeled off the roof. The wind changed direction another 90 degrees and was now coming more from the south-east. In Julie's words, 'It hit with a whoomph.' Julie ran down to her two-year-old son's bedroom and grabbed him from the bed. She recalled what happened next:

 As I was running back down the hallway with Brendon, the whole roof peeled back in one great big tear and then two bedrooms behind me blew away. The fire escape was on the main bedroom that had just blown away, and the front stairs faced directly into the wind facing Doctors Gully. We couldn't go out there, because we could hear all the rubbish from next door hitting the front doorway.

The couple and their son scrambled down the hallway until they reached the lounge room, which was separated from the hallway by a door. Heikki closed the door against the wind, using all of his strength. The family then went into the bathroom, but the manhole cover in the ceiling fell down on Julie's head and cut her scalp open quite badly. Seeking more shelter, they decided to get into the corner where the lounge-room door was and Heikki held it shut with his back. Not long after they had settled there, something fairly substantial came hurtling through the lounge-room door, and the impact blew the firefighter across the hallway and knocked him unconscious. Julie recalled, 'He was out to it. I couldn't see what had happened to him. All I knew was that he was down.'

Still holding her son by one arm, she dragged her unconscious and bleeding husband into the separate toilet and curled his body on his side around the pedestal. She then sat with her son in her arms and braced her legs against the toilet, using her back to keep the door closed against the wind. At this stage, Julie reflected on her predicament. 'I really thought I was going to die—absolutely. The noise of the wind was just unbelievable—it was like a roaring.' Dressed only in a very thin nightie and undies, Julie now felt very cold. Heikki came to, but didn't fully comprehend their situation. His head had been opened with a very deep wound from the top of his skull almost down to the nape of his neck. The bleeding over his wet body made matters look awfully grim.

With a huge force, a gust then blew away the wall between the bathroom and the toilet. Then the wall between the kitchen and the toilet blew away. Julie said, 'I saw the white fridge go flying right past me with the huge Christmas hamper in it that we had won in a raffle!' The Tammos were now precariously positioned in an island of very thin walls on their upper deck, with very little superstructure of the house to support their refuge in the toilet. Julie thought they would be blown out of the toilet and down on to the jagged rubble below in their yard. The toilet door blew off its hinges and on to her back and 'just stayed there for a while'. It seemed like the situation couldn't possibly get any worse, but then it deteriorated further.

 I was hanging on by having my arms around Heikki and Brendon and just gripping on to the toilet base. Then it got real scary, because the toilet door blew away and then not much later the toilet broke off at the base. Heikki kicked it off the toilet floor and over the side of the house. There was only a skinny water pipe left, so we hung on to that. We were now totally exposed with no cover at all . . . Inside, I gave up. I expected to die. I might have even said it.

But they didn't give up. The couple bravely hung on for dear life in the foetal position, with Julie covering her son. The child was dressed only in briefs and was 'shaking like a leaf' from the cold, but he 'didn't offer a whimper'. To the Tammos it seemed like an eternity as they lay on the floor with debris hurtling past them. They were drenched right through as the freezing rain pelted their prone bodies. Then, after another hour, probably about an hour and a half before dawn, the wind started to die. In the dark they took the opportunity to climb down off the floor using the large black steel sewerage pipe as their escape. Julie went first and Heikki handed their son down to her. Once on the ground, Julie slipped on something in the dark and badly lacerated her buttocks. They crawled under some rubble from their own hot water system tank and a wall partition, and used it for cover from the rain and wind. They didn't know if the eye was still to come and didn't want to chance trying to move around in all the rubbish in the dark.

The only thing the Tammos saw while crouching under the rubble was their next-door neighbour flashing a torch around his relatively undamaged house. As soon as it got light, Heikki went next door to check on his neighbour and make sure he was all right. Heikki's head wound had stopped bleeding, but, as Julie described it, 'It was just all white. His head had been cut open right down to the skull bone. He must have a hard head, is all I can say.' Luckily his skull wasn't fractured, even more so because the cut was dangerously close to his spinal cord in his neck. During all of this frightening experience, Julie fretted about her other son, Rohan, who had been admitted to Darwin Hospital during the afternoon and was on the top floor in the children's ward. She worried that he was feeling scared and alone without his parents' support.

At Nightcliff, teenager Lesley Mance had said goodbye earlier in the evening to her policeman brother John as he left for night shift. In the high-set house were her mother, her sister-in-law, Rita, and Rita's young baby. The group had been sitting in the lounge room listening to the tropical cyclone warnings and, as Lesley remembered them, 'those awful siren sounds', when after midnight 'windows started smashing in'. Suddenly, walls started collapsing. Unbeknown to the women, the roof had gone. They ran into the bathroom, as recommended in the cyclone warnings hoping that the extra plumbing and smaller wall sections would provide a stronger refuge. However, the walls there also started to rip off the floor braces and the wind started to suck anything loose out of the room. They had very little with them; even the radio was whipped away in the fierce wind. Soon it was evident that the bathroom was about to disintegrate, and the now terrified women and baby scampered literally for their lives down the hallway and into a bedroom where the double bed was attached to the end wall. 'It all happened very quickly,' Lesley said. The noise was now a roar, and occasionally a shriek, and it was difficult to talk.

They decided to get under the bed. Water was pouring in from where the roof had been ripped, and it was becoming very cold. Lesley's mother, who was solidly built, had trouble squeezing in under the bed. Lesley became very concerned when, for the first time in her life, she saw her mother praying. After only another half-hour or so, the floor beneath the double bed where they were taking shelter gave way. Lesley and her mother fell through the gap and dropped 5 metres on to the muddy ground in the midst of the lethal swirling debris and raging wind. The impact and crash into the rubble cut had caused a wound in Lesley's side that later required stitches, but her mother was now seriously injured with a smashed leg and a long gash that had cut 'right through to the bone'. The noise level was deafening. While Lesley was trying to find her mother in the dark between flashes of lightning, Rita and her baby were still upstairs, clinging to a section of the floorboards that hadn't collapsed. Over the noise of the wind, Lesley could hear her terrified sister-in-law screaming.

Now absolutely frantic, Lesley started looking for help. As she looked around, she could see rubbish and debris hurtling through the air. She had nowhere left to shelter, as the house was literally falling apart. In the incredibly strong gale, with flashing lightning and teeming, wind-driven rain, she managed to crawl next door. Hanging on grimly to the steel railing, she climbed up the stairs of the neighbours' house, where she found them huddled under a mattress. Given that Lesley's house had been demolished and theirs was about to disintegrate, they decided to seek shelter in the neighbours' four-wheel-drive vehicle.

Above the incredible din, Lesley could still hear Rita screaming. 'It was a sound I'll never forget,' she said. Fifteen-year-old Lesley found her injured mother and, with a strength she still cannot believe, carried her to the shelter of their neighbours' vehicle. The pair crowded into the car with the neighbours. Lesley recalled, 'I was terrified. No one spoke in the car. I think we were all in shock.' Not long after they had found refuge in the car, Lesley's sister-in-law fell through the floor still clutching her baby. As she hit the ground, she cut her head open on the jagged rubble. Thunder exploded around the frightened group, making it very difficult to talk—not that anyone had much to say. Lesley continued, 'We saw Rita on the ground and our neighbours bravely ventured out into the storm and recovered her and the baby.' All the clothing the family had on was wet, muddy and torn.

Both Lesley's mother and sister-in-law were bleeding very badly. Lesley said she thought they were all going to die. 'It was like a nightmare . . . I saw Mum praying and I thought, "Oh gee, things are pretty bad!"' The car was now packed with wet, scared and badly hurt people. The wind gusts rocked the car violently, and occasionally debris smashed into one of the body panels. Lesley was worried that the car would be rolled over or lifted off the ground. Rain splattered all types of matter against the windows, but luckily nothing penetrated the vehicle. At around 3.00 a.m. the eye passed overhead and things quietened down for a brief time. The downstairs laundry and storeroom structure protected the car more in the second wind, and although the house above was largely wrecked, it didn't collapse on to the vehicle. Totally exhausted, wet and in shock, everyone fell asleep or dozed off until close to dawn. Lesley thinks her sister-in-law may have been unconscious. She was worried they wouldn't survive in the car, but as she said, 'There was nowhere else to go.' As the storm raged around them, her brother John's entire house was totally demolished except for the steel pylons and a stairway.

Further to the west, in the suburb of Alawa, John and Sue Ryan and their three-month-old son Daniel were nervously waiting for the storm to hit. The high-set, three-bedroom government house was typical of the houses in the northern suburbs. Sue was very tired and decided to take her son into the main bedroom and try to get some sleep before his next feed. She was dressed in a nightie. John was 'prowling on patrol', clad only in stubbies (shorts) and a T-shirt. He was looking out into the gloomy night to try and see what was happening around the house. Just before 11.00 p.m., their son's bedroom window blew in with an almighty bang. The noise had been increasing steadily for the last two hours. The house started to disintegrate very soon afterwards, and the noise level increased ten-fold as the first wind struck hard at their house, which was side-on to the wind. Sue recalled, 'It was impossible to talk to each other.' John added, 'The low-set house next door was lower, and its roof peeled off and cannoned into our house. It fell to bits quickly after that.'

Over a period of about forty minutes the house was totally destroyed. The Ryans were desperately trying to get out of it as it came apart. The wind was now shrieking, and large bangs and thuds complemented the roar of the gale as debris impacted into the deteriorating structure. They were soaked by the rain that poured into the house. Sue was carrying Daniel in a carry-basket as they moved to the bathroom in the hope that this last refuge would protect them from the fury of the storm. She managed to grab a brunch coat to put over her nightgown. To reach the bathroom, they had to move across hard wooden floors strewn with broken glass from the shattered louvres.

They had been in the bathroom for only ten to fifteen minutes when the entire roof went. Unbelievably, the noise increased further. John could see the sky above, passing debris and sheets of lightning. He turned to Sue and said, 'We have to get out of here!' The two end walls of the rectangular house, and the whole roof structure complete with trusses, disappeared in one piece with an enormous explosion. The side walls of the house fell in slightly and collapsed on to the remaining furniture. It was now past midnight, but as John said, 'I wasn't looking at my watch.' Sue said of their time in the bathroom, 'It was fifteen minutes of sheer terror.' They had no option but to try and move downstairs once the bathroom ceiling went.

Now totally drenched, they made their way over and around the rubble of their house, guided by the light provided by 'an astronomical amount of lightning. At times it was almost like daylight.' Now that they could see around them, they were worried about the large amount of debris that was flying past. Sue said, 'It was still blowing, but not as strong as when we crawled down the stairs. We were on our hands and knees and hanging on for dear life.' John was now carrying Daniel. They had to hang on to bits of wall and floor as they crawled across the jarrah floor of their house towards the stairs. It took all of Sue's strength not to get blown away by the wind ripping past them.

It was impossible to look into the wind, and the rain was stinging as it lashed their bodies. Sue's feet had been badly cut upstairs when they had run across the smashed louvres. The couple slowly crawled out on to the landing and then down the concrete stairs using the wall of the house to provide some protection as they headed for their concrete-block storeroom, which provided the nearest shelter. John shuffled along dragging the baby's basket and using his body to shelter the tiny child. They eventually made it into the storeroom. With nothing else left around them, Sue sat on the lawnmower and John slumped on to the cement floor.

Just when they thought they had made it to a safe, strong refuge, the wind pressure built up and the Besser brick wall caved inwards at one end. They had only been in the storeroom for thirty minutes and had just recovered from the strain of fighting their way against the wind. Luckily, the Ryans had placed baby Daniel in a pram that was in the storeroom, for it surely saved his young life when a slab of concrete wall weighing around 300 kilograms fell over the top of him. 'This was really when it was most terrifying, because it all fell in on top of Daniel,' John said. Fearing the worst, they scrabbled frantically in the rubble trying to get the heavy bricks off the baby. Fortunately, fixed to the wall was a large piece of board on which John had hung all his hand tools, and the board reduced the impact to a certain degree. They finally dug Daniel out after three or four desperate minutes. He had cried when the wall fell on him and had some scratches, but he appeared to be okay.

Sue now began to think that they would all die. 'I said a lot of prayers,' she said, 'because we were running out of places to go.' John wasn't as pessimistic, but he was very concerned. 'I thought we were in strife.' With the storeroom now in danger of disintegrating, the cars were their last option. When the eye passed overhead, they took the opportunity to move quickly into the Volkswagen sedan. They spent the rest of the night and the second wind in the car. Fortunately, the remnants of the storeroom walls protected them from the second wind. Every time the wind gusted, the lightweight vehicle would rock violently. But, as John commented, 'At least it was some protection.' It was 'pretty hot and horrible' in the VW, he said, as they had the windows right up to keep out the cold rain. For the next two or more hours, debris continually smashed into the car. Baby Daniel cried occasionally and seemed upset, and slept only fitfully. Sue and John sat and waited, listening to the roar of the tropical cyclone around them. At dawn, their other vehicle—a utility—had its roof caved in.

Sue had been breast-feeding Daniel, but she now found that when he wanted a feed, 'it all dried up and went away'. Sue admitted that she was severely traumatised when the wall fell in. John's main concern was that 'we had played our last card and there was nowhere else to go'. Above them, their house had been totally destroyed. All that remained was the kitchen sink and part of the bathroom toilet. The long side walls of the house were still present, but were twisted and lying on the floor and the remnants of the inner walls. Their furniture was in their backyard—and in their neighbours' yards.

In Lee Point Road in neighbouring Wagaman, Tricia and Graeme Clarke had decided to retire for the evening and 'sleep through the storm'. Graeme had an early start at work on Christmas Day and wanted to get some sleep. They had two very young children, nineteen-month-old Sharon and three-month-old Barry, who had been born prematurely and wasn't long out of hospital. Their plan had been for Tricia and the children to spend Christmas Day with her parents. They had even prepared for the event by placing all of the Christmas presents in the boot and back of their car.

Lying in bed in normal Darwin wet season night attire (in other words, nothing), Tricia and Graeme listened as the wind grew louder and louder. Their bedroom was at the end of the house, with louvres on one side and sliding glass windows on the other. The roar of the wind prevented them from dropping off to sleep. By the time the first wind was upon them, they could feel they were getting wet in the bed from rain being forced through the closed louvres. They moved the bed to avoid the rain, but that didn't work. So, being the house-proud person that she is, Tricia got up and mopped the floors! Water was flooding down the passageway along the side of the house that faced the wind. Deciding to try to stem this liquid intrusion, the couple got dressed in street clothes and started gathering buckets and towels to mop up the water. Graeme went into the next bedroom to check on the kids to make sure they hadn't woken up or were getting wet. It was a most fortunate move. As Tricia said, 'The timing was crucial for us.'

As they were mopping up, Tricia heard over the radio that the centre of the tropical cyclone was going to cross the coast at Shoal Bay at 6.00 a.m. When Graeme asked where the cyclone was, Tricia replied, 'It's just up the road. We're going to get it.' The stark reality of the situation now hit home. As Graeme said, 'It pushed us into action mode.' While Tricia was busily mopping up water, Graeme went down to the end of the house closest to the wind and looked out. He noticed that the sliding windows were flexing an incredible amount with the pressure of the wind. During several large flashes of lightning, he peered out of the kitchen window and couldn't believe what he saw. 'There were definitely several black tornadoes inside the cyclone.' For a moment, he was stunned. He knew that tornadoes didn't happen in the Southern Hemisphere, so he couldn't see how it was possible. In fact, tornadoes are associated with severe tropical cyclones below the Equator, but are only rarely reported.

The sight shook Graeme up, and he went back to Tricia and told her what he had seen. They decided to move everyone to the end of the house to be near the kids and close to the bathroom. With a mighty boom, the sliding doors near the lounge room then exploded, and in one incredible gust the far end of the house disappeared into the inky night. Graeme was momentarily stunned. He recalled, 'I could see the trees across the road in the area that wasn't yet developed.' When the end wall went, the wind got in under the ceiling and the entire roof also disappeared into the night.

Graeme said to Tricia, 'Let's move the kids out of the bedroom and go to the bathroom.' Within seconds of their grabbing the children and leaving the room, the children's bedroom was also torn to pieces.

The Clarkes were now sheltering in the corner between the hallway and the bathroom. Suddenly, a wall with built-in cupboards fell over on top of them. Graeme thought they should now stay under this shelter, because it was providing some protection. He tried to knock a hole through the cupboard side that was over his back, with the intention of putting the children in the linen cupboard if things got really bad. Several weeks later, when his elbow was swollen like a mango and very painful and bruised, he remembered that he had wrecked his elbow trying to break into the linen press. They now felt reasonably safe, the noise of the wind was very loud, they were very wet and had to shout to make themselves heard, but the wall over them offered reasonably good protection. Tricia recalled that her greatest concern at this time was for the children and trying to protect them. They began singing carols and lullabies, praying, and trying to occupy Sharon. Baby Barry, who didn't make a sound throughout the entire ordeal, was drenched through and began to develop pneumonia.

The Clarkes recalled that when the eye arrived, the air became very still. Tricia, a born-and-bred Darwinian, was hoping that the storm was over, but in her heart she knew that it was the eye. Graeme decided to look around briefly with the intention of trying to get downstairs. It was very dark as he crawled out from under the rubble, and he cut his knees on the glass that was scattered over everything. He returned and told Tricia he thought that all that was left of their house was some interior walls, which were leaning crazily against each other; the floorboards and bathroom were also still intact. Unbelievably, all they had been sheltering in was a pile of rubble on top of exposed floorboards and a few smashed and leaning walls. Tricia recalled that the damage seemed excessive, but 'I thought it was just our old house'.

During this lull in the storm, they saw a neighbour moving around in his house using a torch. They crawled back in under their pile of rubble and sat on the floor. They had decided to stay where they were, as there was dangerous debris blocking the exits on the stairways and they had no torch.

The eye 'seemed to last about ten minutes' before they heard the second wind coming. The noise was deafening and made the hair stand up on the back of Graeme's neck. He said, 'It came out of relative silence and was just like a freight train, but ten times noisier than the first wind. It just whipped everything that was covering us off . . . and the bathroom wall went like a sheet of paper in the wind.' Their pile of rubble and cupboard shelter was simply whisked away. They were now totally exposed as they sat on the bare floorboards. Little else remained; every wall and stick of furniture was now gone. In *Tracy* parlance, all they had left was a 'dance floor'. They crawled across the slippery wet floors and huddled—hugging each other and their children—next to the bath, which was the only thing left in what had once been the bathroom. Tricia had filled the bathtub with water, but it was now empty. She remembered that they talked to each other about dying.

Graeme thought that they should climb into the bath, but just as they were about to make a move the entire family was picked up bodily and whisked away into the air. Not only were Tricia and Graeme—still grimly holding their babies—airborne, but so were all the floorboards of the house. Like leaves in the wind, the couple were hurled into black, wet, whirling space. They smashed down on to the ground, but neither one had any idea where they were. It was pitch black, and the driving wind and rain made peering into the dark night almost impossible. They had lost the sense of sight. Amazingly, each was still holding a child; Graeme had Sharon, and Tricia had Barry. The normal reaction when people are airborne is for their arms to fly out and drop whatever they are holding, but they both still had their babies in their arms. They couldn't yell to find each other, because their calls were whipped away in the ferocious gale.

Graeme had landed in the yard next door and was lying on the ground, covering his daughter. He desperately looked around in a circle each time the lightning flashed, to try and determine where he was. Then he saw the wreck of the house and their two cars underneath and began to crawl slowly towards them. He carried Sharon under his body much like a koala carries its young. He was being cut and hit by bricks, pieces of timber, sheets of iron and all manner of rubble. The wind was whipping the muddy ground into a fine dirty spray that made breathing difficult as he crawled desperately into the gale. Finally, he got to the cars and shoved his daughter into the front seat of their Holden sedan.

As the car interior light came on when Graeme opened the door, he could see that young Sharon was matted over her entire body in blood. 'I had to clear her eyes so that she could see. She was totally naked. Her nappy had been sucked off her.' Graeme frantically checked her for wounds, but she wasn't badly injured—she was covered in Graeme's blood. He reached up on to the dashboard and switched on the car headlights and then went crawling out looking for Tricia. She had landed badly in their own backyard. Her shoulder had been deeply gashed. When she came to her senses on her back, she recalled,

 I thought that this was it, when the wall got whipped away. I had small Barry wrapped in my arms. I may have been knocked out . . . I couldn't believe I was still alive . . . I wondered where Graeme was and what I should do next . . . I could hear Graeme yelling out, but he sounded so far away and I wasn't sure if it was real or not. He called out, 'Are you all right?' I answered that

I only had 'half of Sharon', but then I realised I actually had little Barry—and all of him.

Graeme was unbelievably relieved to hear his wife's voice through the roaring gale. He eventually located her in the yard and guided her back to the cars. It was extremely difficult to move and they only managed a metre forward at a time. They looked around when the lightning flashed to see where they should head next in order to avoid the rubble that covered almost every square metre of ground. Graeme recalled:

> It was almost impossible to crawl because we were being belted by debris. We had to crawl over all this tin and sharp stuff with nails sticking out . . . It took us ages to crawl around the debris . . . I was constantly yelling to Tricia to guide her as she followed me back.

Graeme now carried tiny Barry, as Tricia had lost the use of one arm. Eventually they crawled on their hands and knees back to the Holden car where their young daughter Sharon was standing on the front seat, not saying a word. Just opening the car door was an absolute effort; but once inside, the family was safely together again. In the dim light inside the vehicle, they looked at each other. Graeme said of Tricia, 'She was wet and bloody.' Tricia added, 'Not a pretty sight.' Almost all of their clothes were stripped off and they were covered in blood and mud. They urgently needed a bandage for Tricia's badly injured shoulder. Tricia said, 'Oh, no!' when she realised she could see the bone in her upper arm through the deep gash. Graeme grabbed a shirt from the back seat that had been intended as a Christmas present. He ripped it into strips and applied it as a bandage. After rendering first aid to his wife, he climbed into the back seat of the car with Barry. Tricia recalled, 'I was in the front with Sharon, who was standing on the seat—stunned, but quiet and brave.' Wet through, cut, bruised, bloody and muddy, the family rode out the remainder of the second wind in the car until dawn.

Forestry officer Laurie Gwynne and his wife Shirley had lived in their high-set three-bedroom house in the northern suburb of Wagaman for six years. The house was aligned east–west and had all three bedrooms at one end, a small balcony and a store-room/laundry underneath. They owned two cars that were parked underneath the house. One of the cars was a Mini-Moke, which Laurie had parked over the towing frame of a box trailer next to the store-room wall. It was parked next to Shirley's Christmas present—a brand-new Mazda 808 sedan. Shirley wasn't working and was on home duties with daughter Tahnee, who was almost three, and Damian, who had just turned one. Shirley's twenty-eighth birthday was on Boxing Day, but, as she recalled, 'I didn't think I would live to see it.'

By midnight the couple knew that the storm 'was deadly serious'. The house was moving on its piers and starting to break up. Around 1.00 a.m., Laurie said to Shirley that it was 'going to go'. Rainwater was coming through the ceiling and Laurie thought that perhaps it was the solar system leaking. It wasn't. In one fell swoop, their roof had gone, leaving only the plaster ceiling and walls. Everyone was under-dressed for the storm. Laurie was wearing swimming togs and a T-shirt, and Shirley was in a short dress and underwear. Tahnee was in summer pyjamas, and Damian was dressed in an all-in-one baby suit. No one was wearing shoes. Everyone began getting very wet, so Shirley grabbed the kids out of their bedroom, along with some rugs to protect them. Laurie said, 'It was scary. We constantly heard explosions, like when the main bedroom blew away, and stuff ripping.'

The bedroom walls were rapidly torn off the floorboards as the full force of the first wind tore into the structure. Everyone rapidly relocated to the lounge room and huddled close together near the front door of the house in case they needed to get out. Shirley had Damian in her arms. Fortuitously, Laurie grabbed a bullock skin rug that normally adorned the polished floorboards in the lounge room and covered the family with it. Not long after, the glass louvres exploded and shattered, sending shards of glass into the rug.

They could see with a gas lantern the destruction occurring around them. Laurie recalled that he could see the glass bending in the windows. He turned the lantern off, fearing that it might be broken and become a lethal missile or start a fire. The wind was now pulling at the remainder of the structure and twisting and tearing at the remaining walls. All that was left upstairs now was the lounge room, kitchen and bathroom with toilet. Worryingly, the banks of dozens of metal and glass louvres were moving backwards and forwards in the wall frame.

In one huge gust, the lounge furniture was sucked out through the partition of the wall that contained a bank of louvres and flew out over the small front balcony. With a huge hole torn in the wall, the remainder of the house began to break up. It was time to move downstairs. Laurie moved to open the front door and, as he did so, it was sucked from its hinges. Shirley recalled:

> The air pressure was so intense . . . Our black-and-white cat was rapidly sucked out over balcony. Laurie had grabbed the two kids and got down the front stairs . . . At that stage, all that was left on the balcony were the floorboards, railing and the steps . . . I grabbed hold of the railing to avoid getting blown over the side and then found I couldn't let go.

Laurie was now screaming at Shirley to come down the stairs, but in one powerful gust she was blown off the balcony and out into the front yard. 'I was beyond worry,' Laurie said. 'Now we were fighting for our lives.' He reunited the family by crawling out into the yard and feeling around, and eventually found his wife. She was injured from her airborne exit off the balcony and, being out in the direct wind, was being hit by flying concrete blocks. Somehow during the turmoil she had sustained a dozen 100-millimetre nail holes in her feet. Shirley now feared for her life. 'The noise was deafening, terrifying. The rain was so heavy, I was almost drowning . . . I couldn't breathe for the water in my face.' She added wryly, 'This was the best part, because later it got worse.'

As they tried to get back under the house, they were hit by sheets of iron and other debris blowing along the road, Shirley said, 'It was like being in a large washing machine.' Laurie led Shirley back to the shelter of the wall of the storeroom where she was reunited with Damian. The illumination they thought came from lightning flashes was actually from the power lines flashing and whipping through the air. Laurie was now trying to move everyone into the storeroom which would provide some cover, but the direction of the gale made it impossible to reach the storeroom and they had to stay on the leeward side and out in the open—but still under the remnants of the house. They constantly had to move to find cover from the wind, which seemed to change direction. They screamed at each other to let the other know what they were doing. To attempt to get through the door of the storeroom in the full force of the wind was to invite certain death.

Bricks were now flying around from wrecked houses further up the road. Shirley received the brunt of the impact, and the bricks piled up against her back as she sheltered her son with her body. Not far from the storeroom there was an above-ground swimming pool that contained about 40,000 litres of water. The pool collapsed, and Shirley was in the direct path of a wall of water that now poured forth. As she tried to move to a safer spot, her son Damian was washed from her arms. She recalled:

 I thought I had lost him forever. And now I had also lost contact with Laurie . . . Even though I had only moved about 6 feet [2 metres] from someone, it might as well have been 10 kilometres, because I just had no contact at all. So, when Damian went, I panicked, and I can remember screaming in absolute terror . . . He weighed about 15 kilograms and was the biggest, fattest baby, and that was probably why I lost hold of him.

Shirley, now totally distraught, crawled around in the dark on all fours feeling for her baby boy, who had been washed away in the direction of the backyard. She recalled that she kept being hit with things. She used the storeroom wall as a reference point, because she couldn't see anything. She got to a corner of the storeroom after what seemed an eternity and was now on her own in front of her new car. She sat on the ground in front of the car with her back up against the Besser block wall. The Mazda was parked in gear, and after a while the wind changed direction again and, frighteningly, she could feel the car's bumper bar pressing up against her. The wind, measuring 200 kilometres-plus per hour, was now behind the vehicles and pushing one slowly towards her and threatening to crush her against the wall. She changed position to put her foot up against the wheel in an attempt to fend off the car, and as she did so she felt her son Damian's body under the car, wrapped around the tyre, and frantically grabbed him.

Shirley couldn't tell if Damian was alive or dead. She was mortified as she hugged him to her chest. Damian only had a singlet on—his small babysuit had been torn from his body. As she clutched him she felt the car once again nudging closer and threatening to crush her. With a strength that can only be found by someone who fears dreadfully for their life, she pushed the car backwards into the wind with her legs. Shirley Gwynne is only 165 centimetres tall and weighs around 60 kilograms; she is not a strong woman. But her instinct for survival and maternal, protective drive gave her the physical and mental power to push the car off herself and her child. The car went backwards and she moved a few metres to the side. As she did so, she found Laurie, who was sheltering nearby and next to the box trailer.

Getting into the car was simply not an option, because they were unable to cover the couple of metres into the wind; besides, the sedan was locked. The open Mini-Moke offered no protection. Laurie screamed at Shirley to climb into the trailer, whose sides offered minimal cover. Amazingly, their cat must have been blessed with a good homing instinct, because it was also now in the trailer. It was only now that Laurie and Shirley thought they would be all right. Above their heads on a rack was a 4-metre-long dinghy which had broken loose and was in danger of falling down on to the family. As it swung in the wind, Laurie put his foot up and kicked it off the cradle. He said, 'The wind caught the dinghy and it went up into the air like a sheet of paper and was never seen again.'

In the dark, Shirley heard a faint, desperate call for help. She said, 'It sounded like it was miles away. Then, as the wind changed, the voice sounded louder and we realised it was right next to us.' Laurie was only a metre from a woman from next door who was huddling next to the trailer. They didn't know her very well, as she had only recently arrived in the street. She had crawled down her own driveway and into the gutter of Lee Point Road and, using it for the barest of cover, had somehow crawled along the road, up and over the rubble, and up the Gwynnes' driveway. She was desperately looking for help for her husband and baby who were trapped by a collapsed wall in their wrecked house. Her husband was trapped from the hips down under a bed—still holding the baby.

There was no eye of the storm for Laurie to take advantage of to quickly move and conduct a rescue, as this part of Wagaman suburb stayed in gale-force winds the whole time. Laurie recalled, 'I went with this woman. Obviously, I was loath to leave my own family, but you just can't leave people.' By now Laurie had found a pair of boots which he pulled on. He added, 'It took ages to get out of the trailer and crawl back to her house using the gutter for protection.' The wind was so strong it was impossible to stand up, and putting one's head above crawling height was an invitation to have it knocked off by flying debris. At the wrecked house, Laurie searched around in the dark and found a piece of timber with which to lever up the wall and remove the trapped man and baby. Not easily done in the dark, in lashing, teeming rain, and with a raging gale threatening to blow him off the floor. It took all of Laurie's strength just to get into the house and find the trapped man and child. After hauling them from the wreckage, he led his neighbours down the stairs and to the leeward side of their storeroom. Now he had to get back to Shirley and the children. He recalled:

 I made a very bad decision . . . I was only 10 to 15 metres from our cars and Shirley. The wind had dropped from 250 to 249 kilometres per hour and I thought, 'That's a lull', and I decided to make a dash in a straight line for our cars . . . I was thirty-two years old and in my prime, and I thought, 'This is easy. I can do this.' I only made it about 3 metres and a sheet of corrugated iron cleaned me up and swept me off my feet and out into the street— probably 40 metres.

Laurie landed in a pile of rubbish and debris out in front of his house. He was badly cut and bruised from the iron which, luckily, had hit him upright and not like a knife. The impact smashed his toes. Shirley added, 'He was a real mess.' Laurie was upset that he had 'destroyed a good pair of boots'. The air was filled with flying bits of iron, most of which seemed to have come from the nearby Wagaman Primary School. The steel pipe post on the front of their balcony was cut in half at this time. Painfully, Laurie crawled back up their driveway to the trailer and collapsed there until the storm abated. Lying exhausted in the trailer next to his family, with the wind raging around them and debris smashing into the storeroom and their cars, Laurie thought, 'Is this ever going to finish?'

While Laurie was away rescuing the couple next door, Shirley was terrified that the children were dead and began pinching them on the arms to see if they were alive.

> I crouched over, protecting them, and we were still getting hit with stuff . . . I felt their faces, and their eyes were wide open and I thought they were dead. I started pinching their little arms . . . and they would flinch and I then knew they were all right . . . And the next morning they were bruised all over from being pinched.

The family huddled together in the trailer until dawn. Laurie reflected on what might happen to Shirley and the kids. 'It wasn't fear,' he said. 'I was just numb.' Shirley was covered in abrasions, but had nothing broken. Laurie had broken toes and was cut badly in many places and covered in blood. The children were heavily bruised—apart from those that Shirley had inflicted upon them—but no bones were broken. The family spent almost three hours in the box trailer. There was absolutely nothing left of their house, save for the stairs and the balcony railing. Shirley said admiringly of Laurie, 'He was in total control.' But when it was over, Laurie admitted unashamedly, 'I caved in and Shirley bounced back. After dawn, I sat on a heap of rubble and cried.' Yet, this man showed exceptional bravery during the storm, fighting not only for the lives of his own family, but also for his neighbours' lives. They were all dehydrated, freezing cold and candidates for shock. Laurie remarked, 'I went grey in one year.'

In neighbouring Wanguri, Peter and Ann Gray had gone to bed dressed in shorts and T-shirts, 'just in case'. Ann was prepared and had packed all of their personal papers—their insurance policies, wedding certificate, photo albums, passports, and so on—into a plastic bag and had put them next to the bed. Even though the wind was ferocious and the rain was pelting against the louvres, the young married couple weren't too worried—yet. Sometime before midnight they heard their solar hot-water heater being ripped loose off the roof and, not realising the intensity of the storm, Peter chased it as it rolled down the street. He recalled that he had a lot of trouble dodging rubbish and flying debris as he walked back up the street to their house.

About half an hour after midnight, lying in bed, they noticed that the lights went off. They then heard a huge crash. They got up and went down the hallway to where a door was half-open facing into the living room. Peter looked out. He said, 'I realised that I could see the cars downstairs—through the floor! This was when I realised we were in trouble.' Ann recalled that as she was looking over Peter's shoulder into the abyss, 'He said, "Shit!" . . . or something like that.' During the flashes of lightning, the Grays could now see that the far end of their house and floorboards had vanished. The entire kitchen and lounge room had simply gone. The floor joists were still there, but the bearers and floorboards were gone. The bedrooms and bathroom were still intact, and so they decided to retreat to the far end of the house to the master bedroom. From there they tried to get out of the house and down the back stairs, but they were blocked by large amounts of rubble and roofing iron 'wrapped around the railings'. To attempt to leave via the front stairs was asking to be sliced in half by flying sheets of corrugated iron or the occasional passing refrigerator door. Peter said, 'Now the noise was really picking up.' What surprised the Grays was the lack of noise when their house was being ripped apart at the far end. The noise of the rain on the iron roof was deafening, but they thought they would have heard all the floorboards being ripped up.

Their exits from the house were now cut off. It was absolutely pitch black outside, but there was a lot of lightning. The bathroom wasn't a good option, as that side of the house was breaking up. They went back down to the intact master bedroom and sat on the bed. As Ann commented, 'There wasn't much else we could do.' Debris from the nearby Dripstone High School, in Delaware Street, Tiwi, 200 metres away, began slamming into their house. The school had been under construction but was now being demolished by *Tracy*. Sitting on the bed with the door closed, the pair looked up at the ceiling and could see it starting to lift off the wall plates. The corner of the bedroom wall started to split open. Their remaining roof was now very close to taking flight. Fearing some sort of major structural damage was about to occur, they huddled together on the floor on the far side to the wind, Ann still clutching her plastic bag of personal papers.

With an enormous explosion, the entire louvre frame containing three banks of metal and glass louvres blew in on to the bed, just missing the crouching couple. A split-second later, the lee-side louvre frame and windows totally blew out in the wind tunnel of their bedroom. As all of this was occurring, Peter and Ann were blown bodily out of the bedroom and into space, closely followed by their double bed. They landed on the concrete driveway 5 metres below and down from the house and bounced. It was about 25 metres to the closest available cover in the house next door. They remembered that their own storeroom was locked. Peter looked around during the next few seconds of lightning and saw that the house next door was still standing. He grabbed Ann's hand. She recalled, 'Peter yelled, "We'll run for next door!" But we only took three steps and then the roof was on us.' Peter added, '*And* the rest of the house.' The bed had gone over their heads and, thankfully, so had the window banks. The house followed and was now scattered throughout the backyard and on top of Peter and Ann.

They had been 'clobbered by the roof and separated'. Ann may have been knocked out, because the next thing she remembered was coming to and finding she was 'the meat in the sandwich' between two lots of windows. She was unable to move, and was unwilling to try for fear of slashing herself open on broken glass or jagged tin. Peter had been blown sideways another 5 metres and was trapped by his legs under the remainder of the roof and three prefabricated walls. He was on his back, with his foot literally screwed into a window frame. A ceiling fan mast was driven into the mud near his head and was holding the roof off him. In its flight, the roof had actually spun in the air and was back-to-front from where it had lifted off. Peter thinks he may also have been knocked unconscious, as he lost his memory of what happened for a while.

Disoriented, shocked and very dazed, the couple lay under the rubble for what they think was probably only a short period of time. Peter took a while to answer Ann's calls. The storm raged over the top of them, and rain splattered all around into the muddy ground. Ann didn't think she was injured. 'Just sore.' Peter knew he was injured; he had two of the three walls squashing him from hip to toe, and the house roof on the top of that. He tried to move his legs, but he quickly and very painfully discovered that a screw was driven into the top of his foot. He couldn't move. They realised individually that they just had to 'wait it out'. Ann was still yelling out for Peter and desperately wanted him to answer. To her great relief, he finally came to. Ann recalled:

 I did a lot of praying. I really thought I would die . . . so I thought, 'Stuff it,' and I just relaxed . . . I actually thought the army would be there to dig us out at dawn. It was too noisy to talk, because the wind was screaming.

In momentary lulls, Ann would scream at the top of her voice, 'Are you okay?' and Peter would answer back, 'Yes'—and that was the limit of their conversation before the wind picked up again. The Grays thought it was half an hour to an hour before the eye arrived after they had 'left their house'. Now it was dead quiet. There was no rain or lightning. 'It seemed to have eased off.' They were now able to yell to each other from under their respective piles of rubble. Ann thought the storm was over, but Peter knew better. 'Don't move! Stay where you are!' he yelled. Ann couldn't have moved if she had wanted to.

After what they thought was only ten to fifteen minutes, the second wind arrived. They were now really worried about what might happen. Peter didn't think he would die, but he admitted,

'I was worried about the debris when the other half of this wind started. What was going to happen to us?' Lying trapped under the debris, the Grays heard the second blow 'coming up the street and moving things—half a block away'. Ann thought it was 'like an approaching rain squall, but really quick'. Peter described it as being very loud and strong, and agreed with Ann that, for a full minute, they knew it was on its way. They could hear the rubble moving. From where they lay trapped in their piles of debris, they thought the second wind had the same intensity as the first, but Peter added wryly, 'Maybe we were getting used to it.' Again, it was pitch black. Fortunately for this entrapped pair, the second wind didn't affect them, 'but it seemed to last forever'. Peter thought it was about half an hour before first light when the wind abated. In the gloom of her 'window sandwich', Ann looked forward to getting out and having the army there to help with the clean-up. 'We still thought it was only our house,' she said.

During the rampant destruction of Darwin by Cyclone *Tracy*, many people had no choice but to seek shelter in their cars, despite the warnings about using vehicles as shelter. For those whose houses had been torn asunder by the storm, risky cover was better than no cover. Some people were able to stay in one place and survive; others were chased from room to room and from cover to cover. Some sought shelter in a car early in the piece and rode out the fury of the storm. But people like the Clarkes, the Grays and the Gwynnes and their children, who were literally blown out of their shattered dwellings, became the true riders of the storm.

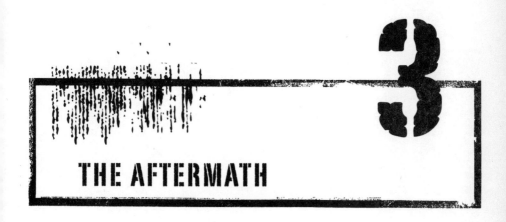

THE AFTERMATH

3

> " It looked like an atomic bomb had hit Darwin. "
>
> — JOHN RYAN

11 THE SOUNDS OF SILENCE

As FIRST LIGHT seeped through the misty, drizzly rain early on Christmas Day, it revealed damage on a scale that was unprecedented in Australia's short history. Cyclone *Tracy* had now moved away from the stricken town and eventually petered out as a severe rain depression well south towards Katherine. Despite being a relatively small tropical cyclone, *Tracy* was extremely intense and produced very strong gale-force winds that flattened the town—and especially the northern suburbs. There was not a house left undamaged, and almost 90 per cent of them were wrecked to a great and uninhabitable degree. Every backyard was littered with debris. In many cases, almost every window, whether it was a louvre or a pane of glass, was broken. It was as if a giant hand had reached down and broken a miniature city by twisting and crushing anything above ground. The roads were virtually impassable unless one had a front-end loader to scrape away the piles of debris that choked them. If there was a clear road, there was a good chance one couldn't get very far before the sharp nails and shards of glass claimed a tyre. And it was eerily hushed. After the roar of the second wind and the tumultuous smashing and tearing of structures, it was now calm and deathly silent. There were no bird noises or cicadas strumming out a whirr—it was just quiet.

The relative humidity was 100 per cent, the ground was sodden from the enormous amount of rainfall that had fallen during the night, and the temperature was rising steadily and sitting on about 22 degrees Celsius at dawn.

On Christmas morning one would normally expect to hear the

gleeful peals of delight as children opened Christmas presents and road-tested new trikes and bicycles out in front of their houses. There were no sounds of merriment on this Christmas Day. Occasionally a neighbour was heard calling the name of a cherished pet; or the sound of a piece of tin being lifted and replaced would reverberate, unmuffled by the tropical foliage that had previously surrounded people's houses.

The vast majority of people who saw night dissolve into dawn on Christmas Day had experienced true and piercing terror for the first time in their lives. They were afraid, in shock and suffering from exposure to varying degrees. Hundreds of them were injured.

In Canberra, Major General Alan Stretton, head of the recently formed National Disaster's Organisation, received a telephone call at 6.25 a.m. from his duty officer informing him that Darwin had been struck by Tropical Cyclone *Tracy*. The wheels of recovery were rapidly set in motion but would take another twenty-four hours to crystallise in the wreckage of this northern city.

The dead that were uncovered and extracted from the rubble were taken to the local police stations or to the hospital. The problem facing the Northern Territory Police was identifying the bodies, as many people had lost their wallets and personal papers. The solution to a lack of morgue facilities was simply to photograph and fingerprint the dead and then bury them in a temporary plot in the Darwin Cemetery.

The high schools and primary schools in the suburbs became the focal points for people to gather and sort out priorities. The basics were simple—food, clothing and shelter. When the radio station began transmitting again, the schools also became the centres for information and evacuation points. Other groups gravitated to places like the local telephone exchange in Wanguri, which was capable of housing quite a few families. People just needed a roof over their heads, and on Christmas Day there weren't too many roofs remaining. Around mid-morning, some enterprising men were able to get front-end loaders on to the roads and begin clearing the debris.

Some people didn't move. They stayed huddled together, unsure of what was happening. Only when they were sure the storm had passed did they dare to stand and look around them. The scene that greeted them on this holy day was catastrophic. The damage was almost obscene in its totality. As Rick Conlon described it, 'Darwin was a 250-square-mile rubbish dump.'

When assistant harbour master Colin Wood eventually got himself back to his house in Schulze Street, not far from the CBD, he found that his wife had been killed. A bedroom wall had fallen on top of her. Colin picked up his wife, put her in the back of his car and took her to the local police station.

Carl Allridge left his partly wrecked house at 6.30 a.m. to try and drive down to the wharf. Within 500 metres he had three flat tyres. He got a lift to the port in a car driven by some tourists who had breezed into town that morning and been dumbstruck by the scene that had greeted them. Carl was so cold he was wearing a leather jerkin under a woollen duffle coat to try and warm his chilled body.

The port was wrecked. There were twenty-nine vessels sunk or wrecked out of about forty that Carl could recall. When the vessel *Nyandra* came in on Christmas Day, the master, Captain Roy Marsh, set up communications. It was a risk tying up to the wharf, because no one knew what was lying at the bottom of the harbour. Carl returned home at around midday to assess his personal devastation. They had lost their end bedroom, and the balcony was now over the roof. The roof had been peeled back and the interior walls had caved into the lounge. The furniture was stripped. The house was basically uninhabitable upstairs, apart from the toilets. The kitchen—apart from the roof—was okay, so Carl's wife cooked their Christmas turkey and later took it down to the wharf to feed the workers.

Around the corner, twelve-year-old Matthew Coffey looked out at the street. There was so much rubble in front of the house and on the road, he thought it would need a rock climber to negotiate it. The stairs of his house were totally blocked by rubbish; they were bent and looked capable of collapsing under the weight, and some of the risers were missing. The only way off the floor downstairs was to get a quilt cover, tie it around the sewerage downpipe in the toilet and shimmy down on to the roof of his sister-in-law's Nissan Patrol, which, fortunately, was parked half under the house. This wasn't too much of a problem for everyone except his mother, who was wearing a dress but no knickers. A large woman, she was embarrassed about coming down to the boys below who were waiting to help. After promising not to look, down she came and they began the task of sorting out the mess.

In the flat he shared with his mother in Mitchell Street, Peter McIver stared across the road; all the houses on the opposite side of the street were gone. Peter knew the people who lived in the old fibro high-set houses and wondered what had happened to them. In the flats at the rear, Block 2 had lost half of the top floor. There was debris everywhere. Peter had a friend living in Block 1 and went around to see him. He nearly stopped in his tracks when he saw that half of the second floor and the whole of the top floor of the three-storey building was gone. He found his mate hiding in a bathroom. As Peter recalled, 'He was totally petrified.' He took his mate back to his own flat and gave him shelter.

Eddie and Greta Quong were climbing out from under the rubble in their inner-city house in Shepherd Street. Eddie noticed that all the water lying around was very salty. He recalled, 'For months, nothing grew because of the salt.' Greta remembered a fellow riding around the streets in town on a pushbike calling out, 'Merry Christmas.' She added, 'It was a frightening scene. No trees, nothing green was left, even the grass was gone.' Eddie said, 'There were no birds, no flies. Our front stairs were gone and the back stairs were full of rubbish.' In their driveway was a piano smashed to bits. A 6-metre fibreglass boat lay strewn across their car—it was 500 metres to the nearest water! Their two sons who had sheltered in the flat downstairs were okay, but they had been very lucky as a star picket had pierced their room. A sheet of corrugated iron was driven so hard into the ground it protruded only 15 centimetres out of the dirt and had to be pulled out with a jeep. The boys couldn't get out the door of the now roof-less flat. All of the internal walls in the house had been knocked out of plumb—but, apart from being drenched and cold, the Quong family were uninjured. They lived in their car for three days, during which time their pet cat returned home. Eddie's cousin, Arthur Fong-Lim, lived at Fannie Bay. Eddie recalled:

> During the storm he went down his stairs and a fridge killed him. He had an 18-cubic foot fridge fall on him during the eye or shortly after. His wife had two boys under a table and got on top of them, but she lost two legs cut off with a sheet of iron.

The Quongs' above-ground pool was reasonably okay, so Greta washed the turkey she had bought for Christmas lunch in the pool water. Eddie's mother had a gas barbecue, and so they made turkey stew and fed their neighbours, 'who all chucked in whatever they could'. Eddie knew several other people who had suffered greatly during the storm, including Malina Bell, the wife of the local member of the Legislative Council, Tom Bell—she perished after being struck by flying roofing iron.

In the CBD, the police were frantically trying to establish local communications. New cars were being requisitioned and taken out of car yards, and long queues were starting to form in front of the city police station. The officers on duty were trying to placate the many distressed and disoriented survivors. Sergeant John Woodcock had managed to drive into town to make sure he had a working shift on duty and stayed there until 2.00 p.m. before returning home. He recalled that medical supplies were urgently needed, and so he simply requisitioned stores and commandeered and signed for them from local pharmacies.

When *Tracy*'s toll on Darwin's maritime community was finally reckoned, Rob Perkins was able to consider himself lucky, though he never went back to sea. He had originally been rostered for duty on the *Flood Bird* but his place had been taken by another trawlerman. All hands were lost on the *Flood Bird*. Six men drowned, and the body of the crewman who had taken Rob's place wasn't found for eighteen months. Another trawler, the *Frigate Bird*, overturned and was stuck in a sandbank on Weed Reef about 5 kilometres south-west of the wharves. Two crewmen were trapped inside the hull but eventually escaped before the next tide came in. The *Lady Cynthia*, an oil rig tender that arrived back in Darwin on Christmas Day, searched the harbour for survivors and rescued two men from the upturned *Frigate Bird*. Two other crewmen from the trawler had perished.

When the crew of the *Clipper Bird* clambered up on to the tilted deck at first light, they were surprised to find that they had been driven ashore at Doctors Gully. Even more astonishing, only 50 metres away, one of the Attack Class patrol boats was in the same predicament. Rob Perkins recalled, 'The patrol boat looked okay except for a couple of holes in the hull below the water line. Beyond that was another trawler that looked to be about 20 metres in length.' Rob wasn't sure whose it was, as the name wasn't visible, but it's likely it was the *Jenny Wright*, which belonged to another company.

It was now low tide, and the *Clipper Bird* was very high and dry. They wouldn't be going anywhere for quite a while. The skipper was on the ship's radio trying to contact other boats. He sent Rob and a few other crewmen ashore, as Rob remarked wryly, 'to get a tug for the next high tide to pull us back into the water'. They threw a rope ladder over the side and clambered down into the glutinous, calf-deep mud and trudged ashore. The men were shocked when they saw the state of the town. Rob said, 'It was an incredible sight. Everything was totally blown away.'

A tug —one of the very few operating vessels left in Darwin— pulled the trawler off the rocks at the next high tide that afternoon; luckily, it didn't need much pulling. The *Clipper Bird* steamed around to the smashed wharves and tied up on the inside of Stokes Hill Wharf. Rob said, 'Beside us, there was a boat underneath the wharf. We spent the rest of the day searching along the foreshores for survivors from the *Flood Bird*. The *Frigate Bird* wasn't found until next day.' Rob recalled feeling a great sense of loss; he knew many of the Japanese and Australian crewmen who had perished.

The *Cat Bird* and the *Clipper Bird* had come through fairly well. The trawler *Bell Bird* had had her rear deck ripped out. The trawler *Blue Bird* had sunk inside Iron Ore Wharf, so only two of the six boats in their fleet had survived. Rob said, 'I was lucky . . .'

A ham radio operator made contact with the *Alanna Fay* on Christmas Day and Laurie Coffey's family learned that he was all right. As the barge steamed into Frances Bay on that fateful Christmas Day, the crew and young Philip Grice could only stand on the aft wheelhouse deck and stare at the damage in the port. Laurie recalled:

 We finally got into Darwin in the morning, and we got a great shock. Here were all the boats floating upside down. Little or no trees. It was a great shock to come into the place and find it under these conditions. I mean, everybody on board was lucky. There was no loss to anybody—apart from houses. Everybody's house was either partly gone, or gone. But there was no loss of life to anybody in the company at all. But as I say, it was a great shock.

When the *Alanna Fay* berthed at the Barge Express wharf, the owner, John Grice, was waiting. He had been in touch with the crew's families and had told everyone what was going on. The crew couldn't drive back home because the roads were cluttered with debris. Frances Bay was also in a mess. Laurie Coffey said, 'The whole waterfront was all snarled up.' Laurie was in for an even bigger shock when he finally walked into his street. His own house was, in his words, 'I suppose more than half gone. All the houses in the street were either gone, or partly gone.'

In the comparative safety of the open ocean, HMAS *Assail* had been able to ride out the storm. When the wind eased to 'around 40 to 50 knots', they reversed course and eventually returned to the wrecked harbour at 6.00 a.m. on Christmas Day. When they had left the evening before, there were some fifty craft of various shapes and sizes in the harbour. But as the *Assail* limped back into Darwin, there were only five craft remaining: 'Two patrol boats, two trawlers and one tug. The rest were sunk, missing or aground.' Chris Cleveland, who at that time was totally oblivious to the extent of the damage to Darwin and the harbour area, recalled his initial thoughts. 'I thought I was the only one who had screwed up. I was wondering how to tell my boss that I had lost a life raft and broken my anchor cable. I had the impression that everyone was still happily sitting on their moorings.' They weren't. In fact, the *Assail* was the only patrol boat remaining from the fleet that was able to conduct search and rescue duties on Boxing Day. In the morning light, Chris could see that his arms were covered in huge bruises from where they had been wrapped around the binnacle as he tried to stay upright in the wheelhouse. He believed that his boat and crew survived their incredible ordeal because: 'I had an excellent engineering staff and all of my equipment worked perfectly. My crew were magnificent, particularly the engineers and electrician.'

When the prawn trawler the N.R. *Anson*, which had been unable to respond to the two distress signals it had seen during the night, limped back to the still intact main wharf at 11.40 a.m. on Christmas Day, the crew couldn't believe the damage they saw ashore. Bob Hedditch said that, given what they had been through, 'Nothing will worry me again.'

According to harbour master Carl Allridge's report dated 4 January 1975, 'At least 29 vessels were sunk or wrecked, several were driven ashore and later re-floated and at least 20 persons were lost.'

At Larrakeyah Army Barracks, 'It was an absolute shambles,' Peter Kerntke recalled. 'The trunks of trees still stood, but with no leaves and branches. The steel telephone poles were bent and twisted like a pretzel.' Peter went looking for his house, but because all of the normal reference points were gone, he ended up counting down the married quarters and found to his utter disbelief that his quarter had been smashed to pieces. If he hadn't insisted on his wife moving to the Officers' Mess for shelter, she and the children may well have ended up down in the sea in Doctors Gully. All that remained of the house was the floorboards and a toilet—still with the Christmas tree in the bowl where Diane had shoved it—and a buffet with some liquor bottles still standing underneath. Debris was piled 2 metres high everywhere in the yard. In the kitchen debris, he found that their saucepans had been sandblasted down to the bare metal. Three months later, a stranger found the Kerntkes' Commonwealth Bank passbook 15 kilometres from their house.

After his family, Peter Kerntke's greatest concern was to establish communications out of Darwin. At around 11.00 a.m., he finally got through to Mt Isa, in Queensland, using a microwave link through the ABC studios in town. They were asked to ring the duty officer of 104 Signal Squadron in Townsville, and a Morse link was up and running shortly thereafter. The emergency plan was to have local communications in Darwin with the police, navy, RAAF and essential services.

Everywhere in Darwin on that Christmas morning, people were simply astonished by what they had experienced. In recollecting the events, they used words such as 'unbelievable', 'stunned', 'numb'. Many said they 'thanked God'. The sheer force of the wind had made a mockery of cyclone bolts in houses; in many cases, all that was left was a string of twisted cyclone bolts pointing skywards. Many people had thought during the night that theirs was the only house being damaged so severely by the cyclone. They cursed builders, their old home or just bad luck. But when dawn arrived, they saw that almost everyone was in the same predicament.

In Larrakeyah, Julie and Heikki Tammo looked up at their house and found they were left with just the remnants of the bathroom. Heikki's badly cut head needed stitching. They stood in their yard and, in the quiet of the dawn, heard a neighbour calling for help. Heikki went across and helped a couple who had climbed into a wardrobe when their house broke up. But a wall had collapsed and the husband, wife and dog, with a litter of puppies, were all trapped—unhurt—underneath half a tonne of rubble.

Trying to get Heikki to Darwin Hospital was a drama in itself. The steel telephone pole out in front of their house was bent over and blocked their driveway, and they couldn't get their car out of the yard. Julie managed to get them a lift from a neighbour in a four-wheel drive, who drove on flat tyres the 3 kilometres to the hospital.

The scene at the hospital was akin to a casualty clearing station from the Crimean War. The emergency and casualty ward had about 100 millimetres of water on the floor and was literally red with blood. The people who had been brought in were in shock. Many were bewildered or dazed. Julie Tammo brought Heikki into the casualty ward not realising that her wet cotton nightie made her appear as naked as some of the people sitting on the floor in the water. Julie recalled looking around her:

> Everyone in there was wet and cut . . . Most people were silent, some were sobbing quietly and others were shaking. Occasionally a voice would cry out in pain. There were at least fifty people there and it was still very early.

Getting sewn up isn't pleasant at the best of times, but now it was worse because the hospital had very quickly run out of local anaesthetic and stitching was being done with only care and soothing words to relieve the pain. There was no time to shave Heikki's head; the stitches went in—hair and all. There were simply too many people needing attention to take the time to prepare the injured for normal first aid and repair. Julie also had her bottom stitched where she had fallen on jagged tin. She was anxious about her young son who had been admitted to hospital the day before, as it was obvious the children's ward had been extensively damaged. Her worries were alleviated when she learned that all the children had been moved down from the top floors and put three to a bed downstairs and that 'he was fine'.

While waiting for her husband to be stitched, Julie and another woman started making urns of hot tea and dispensing it from a tea trolley. As she was handing out cups of tea, she looked through the front door and saw a dead body being brought in. 'It was naked and blue,' she said. Three hours after they had arrived, the Tammos went home, where Julie found her Christmas ham in the yard embedded with gravel. She washed it and then went down to the fire station to hand out food, tell them about Heikki and use the telephone to call her parents. Her brother had driven out of Darwin the previous night when he heard the tropical cyclone warnings and headed off down The Track. He returned the next morning to his house in Rapid Creek to find nothing—not even the floorboards—left of his house.

Grant Tambling's house in Kahlan survived because of its good, solid construction and fortuitous position in the lee of a small hill. It wasn't very far from where Jenny Wood was killed. The wind seemed to be fickle in its destruction; every now and then, a lone house would be left standing among twenty totally destroyed structures. The Tamblings would soon be host to a cluster of people who were staying in Darwin on essential tasks and needed a roof over their heads.

At the airport the devastation was incredible. John Auld looked out from the office where he had sheltered all night and took in a scene of utter devastation and heartbreak. Every single plane that had been anchored on the aprons and parking areas had been wrecked. He said, 'There were aircraft piled up at one end of the strip. There were no leaves on the trees. I recall dozens of aircraft impaled into each other, and the ground was littered with bits and pieces of aircraft.' The planes were usually tied down with nylon ropes, but they had no chance against the fury of *Tracy*. Aircraft were upside down, on their nose, bent, smashed and wingless. The long grass on the side of the tarmac strips had been ripped out of the ground, leaving red–brown-coloured mud instead of lush green grass. International flights would be stopped for almost nine months, but the domestic services would be operating—if not on a commercial basis—by the next day.

When Bill Gough arrived to inspect the damage to the Con Air facility, the workmen had to use tractors to clear the wrecked aircraft to allow access to some of the hangars. Bill used the radio in the one and only aircraft that had survived in the hangar to contact Katherine to let people know what had happened in Darwin. Once aircraft started arriving in Darwin, the pilots had to use visual recognition rules, as there was no radar to warn aircraft of each other's location. The storm hit Bill's company hard; one maintenance worker was killed in his flat and another lost his wife. The airport quarantine incinerator used for destroying illegally imported foodstuffs and other goods was a large affair with a prominent reinforced concrete chimney. Bill Gough remembered, 'It had a

2-metre star picket driven through it like a toothpick through a cocktail onion.'

In the adjacent married quarter patch on the RAAF base, Judith Watson had the same thought as many other people when they took in the kaleidoscope of wreckage around them: 'We are the only ones left alive.' She recalled the water running through her house as if a fire hose had been left turned on in the hallway. In the thick, still morning air, she thought this would be as bad as it could get. 'We had nothing. No basic necessities of life were left. No food, shelter or clothing.' But it got worse, because Judith also had no toilet. The only available pedestal was sitting up on top of their floorboards—but unfortunately, there were no walls left to provide a modicum of privacy. She hung on until that night, when, after dark, she managed to find a tree to hide behind—only to be caught in the glare of an RAAF Landrover's headlights as airmen on duty were checking for looters.

The Watsons had come through relatively unscathed. Judith said that she felt 'sort of guilty, because we hadn't had to run and hide like many others who were in real danger'. Judith's neighbour, Josephine Foreman, was thanking her lucky stars—and God—when, at first light, she was staring at a 10 x 10-centimetre floor joist impaled through the wall of the shower recess in the bathroom just next to where they had been sheltering. Josephine spent some time looking after a neighbour who was hysterical because she couldn't find her son; they finally found him in a wardrobe. The woman's husband was in shock.

In Rapid Creek, Chris Kingston-Lee had taken nearly half an hour to get out of the room where his family was sheltering, as timber and their neighbour's refrigerator were blocking the door and keeping it firmly shut. He recalled his first impressions:

 I thought, 'Holy shit.' I just stood there in silence—I was totally numbed. All of the houses across the road that were high-sets . . . were just dance floors. Power cables were down everywhere and strewn around like a giant cobweb . . . A bloke with a sheet wrapped around him came looking for some clothes. He had no pants because he had crapped in his shorts during the storm, being too afraid to get up and go to the toilet.

Chris was moving around soon after 8.00 a.m. Because he had first aid training and a medical kit, he soon found he was in high demand. Once out of his yard, he looked around in disbelief. In his typical laconic and understated way, he thought, 'This will take a bit of cleaning up.' His mate Dennis Lugg arrived and, for a man who never swore, he stunned Chris with his stream of profanity regarding the storm and the damage it had caused to Darwin. As Chris and Dennis moved down the street attending to casualties of all types, they came across some dreadful sights. In one dwelling, a woman had been hiding in a toilet during the storm when she had been impaled and killed with a large piece of flying lumber. In all, they treated about thirty casualties for everything from cuts and abrasions to fractures. Returning home, Chris looked at the rubbish in his yard. He recalled:

> When we cleaned it up there was about 7 tonnes in the yard. My EH Holden car was squashed flat, and my catamaran had a piece of timber straight through both hulls. In the kitchen the condiment canisters were all full of water with the lids still on!

Chris' greatest shock would come when he went into Stuart Park to his motorcycle shop and found total devastation. His bikes were scattered all over the street, dozens of helmets and expensive leather accessories were gone, and all of his business papers were somewhere south of Darwin. He spent an hour trying to recover bits and pieces and then realised he had nowhere to store anything. Insurance would cover only so much. Chris was totally ruined by *Tracy*. After paying off all of his bills after the storm, the business never fully recovered. While Chris was picking up the remains of his stock, he recalled, 'A bloke drove up on a motorbike and asked if I was the owner. And then he asked me for a trade-in on a 750cc bike! I walked up, got close to his face and said quietly, "Fuck off."'

In Alawa, Helen Auld found all her kitchen saucepans in a cupboard (which had no back on it) with the lids still on the saucepans; they were all full of water. She looked in awe at her baby son's cot, which had a piece of timber driven through the mattress and into the floor. Her husband John was driving back from night shift at the airport. Like so many others out checking on friends and relatives, he found that not only were the roads choked with debris, but he also couldn't find any legible street signs to tell him where he was. The suburb was totally unrecognisable. All along Bagot Road, the steel telephone and power poles were bent at right angles about 30 centimetres off the ground. From talking with some of the neighbours, the Aulds soon realised there was no one to help them. Once that realisation had sunk in, people began organising themselves and helping others who were injured or in shock.

John and Sue Ryan looked out from their car, where they had sought their last available shelter when their house disintegrated; they couldn't believe their eyes. John recalled, 'It looked like an atomic bomb had hit Darwin. There wasn't a square metre without rubbish.' There was a lighter moment of black humour when he came upon a neighbour who had come home drunk from a Christmas Eve party the night before, fallen asleep in a lounge-room chair and hadn't woken up until morning. He was pinned under the roof, which had collapsed on him and broken a leg! John went back to Sue, who was now beginning to feel the pain of a considerable amount of glass impaled in her feet. They had lost everything, with the sole exception of a bar they had bought in Florence. It had been protected by a wall that had collapsed on it. They relocated to Alawa School for shelter. John went back home during the day and buried all the perishable food they couldn't eat in a hole in the soft ground in the yard. He then took all the tinned food back to the school, where the initial thirty or forty survivors had now grown to a throng of about one hundred and fifty people. When it started to rain, Sue held young Daniel out in the rain to wash him, as there was no tap water.

Kevin and Karen Jurek in Jingili were very worried about their baby, who had turned blue with cold. The neighbours gave the child some brandy and warmed her up with dry clothes. The Jureks' low-set, pre-fabricated cement-walled house had stood up relatively well, despite losing the roof and most of the trusses. It was now just a shell with no windows, but the solid cement walls had protected them from being injured. A man walked into their yard in his underpants looking for some trousers to borrow—he had nothing else to his name. Like many people who had pet cats, the Jureks' felines had 'gone nuts' and were never the same again. Their brand-new colour television set was full of water. Kevin picked it up, emptied out several litres of water and, amazingly, it continued to operate for another ten years. There was no door on their fridge, and all of the Christmas food was embedded with shards of glass.

John and Janice Woodcock in Nakara were trying to calm John's totally distraught parents who had lost all of their worldly possessions when their caravan was rolled across the backyard. The van was penetrated in many places by what would have been lethal missiles. John's role as a police sergeant was a consideration as he sorted out his family. After digging his police utility out from under piles of rubble, he found that none of his vehicles would start because the batteries were flat. After some time he tried again and got the police ute going. He drove into the city with a badly injured neighbour whose knee was smashed. On the way into town, he collected another serious casualty whose arm was almost severed. He got as far as the local Casuarina police station and found the officers had established a makeshift morgue in the rear of the station. By 9.00 a.m. they already had six bodies. In all, about thirty people would be found dead in the northern suburbs.

What was normally a twenty-minute drive into town took John Woodcock almost two-and-a-half hours. He recalled thinking, as he made his way carefully along the single car track which previous vehicles had made through the rubbish, that 'hundreds of people would have been killed. The devastation was just so complete.' After checking on the shift at the station, he headed back towards home and saw a couple of Aborigines sitting outside a local hotel. John recalled wryly, 'They wanted to know when the pub would open . . . They've got a good sense of humour.' Later that morning, John's elderly father would also become a casualty when he slipped on wet corrugated iron and badly gashed his leg. He wouldn't be the only survivor to fall foul of the piles of rubble.

In Wanguri, the Charringtons were likening the scene around them to Hiroshima. Their neighbours, the Buchanans, were under piles of storeroom blockwork downstairs; all were badly bruised but otherwise okay. Their car had been flattened by the floor joists collapsing on it. A large piece of lumber had been driven through the car from end to end; anyone inside would have been killed. Rowan Charrington recalled that, 'At dawn there was uncertainty about what to do next. All the power lines were down. We were hit with a sort of numbness. Marcia and I were both very tired and we just wanted to lay down and have a sleep.' A young couple who were neighbours had climbed into their Landrover vehicle during the storm and it had been blown over on to its side.

In all this incredible wreckage, amazing stories began to unfold. Houses had been totally wrecked, yet a Hills Hoist clothesline was intact with, albeit fairly grubby, washing still on the line. Judith Watson recalled that the phone table in their hallway, which had a papier-mâché-covered Milo tin on top for pencils, 'went down the hall intact and stayed there! The tin didn't blow away . . . but the rest of the house did!' Her German neighbour had prepared about a dozen small pullets for Christmas. With a grin on her face, she asked Judith to come and look at them. They were stuck up on the ceiling on the ends of the broken rafters in her house like edible Yuletide decorations. Back in her own wrecked house, their cat was in the remains of the roof and couldn't be enticed down.

Peter and Ann Gray, who were trapped under their roof and other house rubble in Wanguri, had to wait until first light before they could attempt to extract themselves. Ann had to take care not to cut herself on the glass in the louvres that pinned her to the ground. It took her twenty minutes to wriggle out from underneath—still clutching her plastic bag of personal papers—and then she and a neighbour from across the road, Terry Brian, went to Peter's aid. Terry offered Ann a top to wear, as little remained of the one she had been wearing. Peter was badly crushed under the weight of the roof and had a large screw through his foot. When Ann tried to lift the roof off his leg, the pain was excruciating, and so they decided to try and jack or lever the walls up instead. The main problem was that the ground was so sodden from the rain that anything they used to try and lift the rubble simply sank back into the mud. Eventually, a platform of bricks was used to provide a firm base, and after thirty minutes of struggle they were able to get Peter out. He recalled looking up at the house that he and Ann had been blown out of.

'The only thing left was about a quarter of the floorboards and a heavy cupboard full of tinned food that had fallen over and not been blown away.'

Once Peter was freed, their helpful neighbour Terry Brian asked if they could come and help him get his wife out of their caravan. They had been building their own home and living on the site in a van while the house was under construction. The Grays accompanied Terry around behind the wrecked house where, to their astonishment, they saw the caravan leaning over and a half-cabin 5-metre boat impaled half-way through its side. Ann feared the worst and didn't want to see a dead body, but Terry assured her that his wife Jackie and their twin seven-year-old girls were okay. A mattress had protected them as they huddled under the table. Terry's problem was that his wife was simply too scared to come out of the van.

Despite their extensive lacerations and bruising—and an unsuspected broken tailbone—Peter and Terry hobbled off looking to help others. They went around the corner into Henbury Avenue and found a neighbour, George Parker, in the back corner of his yard. Peter said:

> He was still in his pyjamas and half-buried under a pile of rubble near his barbecue. He was a dreadful white colour and very cold. He moaned when we moved him into a passing station wagon. He died at Casuarina police station.

George Parker's wife Madeline was almost scalped, but survived. A brother and sister at the same address both had broken limbs. Peter and Terry helped move the injured family to a neighbour's house and tried to comfort them. Like the Grays, they had all been blown out of their house, but they had been hit by flying debris as they had no protection when they hit the ground. The yard was about a metre deep in rubble. It wasn't until the next day, when Peter started to really feel pain, that he realised he also had broken ribs. Keeping busy was good for him, he reflected. He recalled seeing a front-end loader go past his house carrying three casualties in the bucket on the way to somewhere. Eventually, the Grays were advised to head to Wanguri School. When they arrived there in the mid-afternoon, things were being organised. Ann recalled, 'A natural leader had gripped the group up, and they were boiling water and sharing out food and clothing.'

Tricia and Graeme Clarke, who had also been blown around their backyard, must have looked a mess. When Graeme walked up the street to get help from a mate, he recalled, 'Bob told me that he saw this old man walking towards him, and it wasn't until I was 2 metres away that he realised it was me.' Bob's house was a low-set brick construction, and all that remained were low walls only a metre high. They had survived under their kitchen table. Graeme desperately needed to get Tricia to hospital, as her badly cut shoulder was in need of urgent attention. A neighbour was able to get his four-wheel drive out from under the rubble of his house and drove Graeme back to collect Tricia and the children. Within 500 metres all the tyres were flat, but they continued on towards town driving on the wheel rims. Somewhere along the way they were transferred to a police paddy wagon and eventually arrived at the hospital. Tricia remembered the ankle-deep red water in the casualty ward, where she was stitched up by an army doctor who apologised for the lack of anaesthetic and his poor needlework. Their ailing son Barry was admitted to hospital with pneumonia.

Totally exhausted from their ordeal, the Clarkes walked out of casualty and realised they had nowhere to go and no car to go anywhere in. Tricia's watch had been ripped off in the wind, but they think it was about mid-morning when they walked down the road into the CBD to go to Tricia's sister's flat and bunk down there. They trudged down the road towards town, still feeling the cold and clad in little but a blanket. Tricia's sister Sue had managed to get out to their house in Wagaman, found their car covered in blood—but no Clarkes—and feared the worst. She returned to her inner-city flat at 3.00 p.m. to find Tricia and Graeme asleep in her bed. The Clarkes' house was totally destroyed and they were left with absolutely nothing. Tricia recalled, 'Our fridge was two blocks of land away. And several days later, we found one of our suitcases down the other end of Lee Point Road about 2 kilometres away.'

Laurie and Shirley Gwynne climbed out of the box trailer from under their totally destroyed house in Lee Point Road. Shirley was turning black and blue all over her back and sides from where bricks and concrete blocks had struck her while she was crawling around trying to find her son. Her back was so stiff from when she had landed in the yard after being blown off the house balcony that she could hardly move; once she got into a position, it was difficult to move again. She had red streaks running up her legs from the nail holes in her feet and ankles—a sure sign that tetanus was developing. Laurie's toes were a broken mess. He was dressed casually in a pair of swimming togs, but grabbed a stray coat that was lying around. He looked magnificent, he said, in torn work boots, togs and a dinner jacket. The children, who had been so quiet during the storm, now began crying and screaming. The combination of wind, rain, low air pressure and cool air had made everyone very cold. Laurie found the keys to the car in Shirley's handbag in the backyard and started the car. Everyone then sat inside with the heater on full blast to try and warm up. The Gwynnes were left with virtually nothing except for their cars and the box trailer. Some friends, the Blythes, came around, and Shirley started smoking again. Laurie scouted around and found their fridge a few hundred metres down the road with baby Damian's milk bottles still inside. He recovered them and some other basic food items. The turkey that Shirley had prepared was down by the front gate. As Laurie said, 'It had flown without any feathers.' They removed the gravel-impregnated skin from the cooked bird, washed it, and ate it with their friends and neighbours.

People were just sitting quietly in the rubble of their houses. Occasionally, someone was heard calling out to a mate or a neighbour to see if they were okay. Word came to move to the local Wagaman School, about a kilometre away. The Gwynnes grabbed some blankets and set off slowly with their neighbours. Laurie had to hobble along using a hockey stick as a crutch. Shirley recalled:

 Most of Wagaman was there. We got into the school, which still had its roof, and we stayed the night. I sat there all night wide-awake. It was very uncomfortable. People were in shock, crying, yelling at each other and sometimes screaming. Laurie sat watch with a neighbour, Ken Hunter.

The personal dramas unfolding in the school emergency centres weren't conducive to rest and recuperation. When a friend with a truck called around to check on the Gwynnes, they headed off to the forestry research station at Berrimah, which was relatively intact, and stayed there until they were evacuated several days later.

Chris Collins, in Nightcliff, spent some time helping his neighbours to get out from the rubble. He then jogged over to his wife's parents' place 500 metres away. He was stunned by the total devastation he saw all around him. The water tower on Pandanus Street had a fridge indentation in it a hundred metres or so above the ground. His parents were safe, despite having ridden the bathroom's suspended concrete slab on to the ground in the corner of the yard when the house collapsed during the second wind. Chris said that when he returned home, he realised, 'When everything is wiped out, the only people who can help you is yourselves.' At home, he took stock. All that was left of the low-set brick house were the corners that were half a metre or more high and the interior walls. He added, 'Our car was okay, except that there was a verandah parked on it. The tarpaulin was still over the bed, but now it was held down by about a thousand bricks.'

What was also amazing was the volume of vegetation that had disappeared from Darwin. Hundreds of thousands of tonnes of leaves, branches, plants and shrubs had simply disappeared. When Lesley Mance and her family, who had fallen through the floorboards, got out of their neighbours' car, she saw her policeman brother John walking around their totally destroyed house. A look of profound relief crossed his face when he saw his family picking their way towards him through the rubble in their neighbours' yard. Lesley's sister and mother needed medical treatment, so they were loaded into the police utility John was driving and they headed into town. Lesley recalled the horror of that trip:

 There were other people in the back. One had a tarpaulin over him and the floor of the tray was awash with blood. When we got to the casualty ward, the hallway was full of people—some were sitting on the floor. A nurse took Mum away, and next day she was evacuated. A rumour started while we were waiting that the cyclone was coming back, and this freaked me out.

Rick Conlon arrived at the hospital around mid-morning. He described the casualty ward as 'a living hell'. There were sixteen wards in the hospital, but after *Tracy* only two-and-a-half wards were left habitable. The walls of the wards were cavity brick and had stood up fairly well, except for the end walls in the children's ward, and most of the roof was missing. The main problem Rick now faced was the mortuary and refrigeration. He and some colleagues solved the problem by 'liberating' some freezers and plant. Out at Howard Springs, Rick's wife found that the best way to get rid of all the water in the house was to drill holes in the pine floorboards.

Peter Coombe spent his morning using his rental cars to drive around and check on his many friends. He went down to the sailing club where a trimaran was on the roof of the clubhouse and another boat was in the bar; there was beach sand everywhere in the building. The cars in the car park had all been sandblasted down to the silver metal on one side, and a car with a broken window facing the sea was filled with sand up to the windows. The Pickerings, who had been sheltering with Peter Coombe at their friends' house in Fannie Bay, had a communal breakfast washed down with brandy. Judy recalled, 'It was very quiet. People were subdued. Everyone was exhausted from lack of sleep.' Tony Pickering walked back to their house in Ludmilla, since it was impossible to drive anywhere. The house was relatively okay, being in the lee of a small hill, but their above-ground pool was now 500 metres away in a local park—still intact, but minus the water. Under the house, Tony's 1962 model Holden sedan had been turned 90 degrees between the pylons and was unable to be extracted without being physically lifted and turned. A big traveller's palm in the front yard had a roofing member 2 metres long driven a metre into its trunk; the impact had pushed the tree over to one side.

Kids will be kids. Just up the road from the Pickerings, Alan Grove's children wanted to know if Santa had been and brought them presents. The toys were 'floating around in the lounge', but despite much exterior damage the house was okay. What Alan and his wife found difficult was 'the sensory overload of all that damage'. His close neighbours gathered and ate some Christmas cake together at around midday. They had a Christmas lunch cooked on a barbecue. Alan added, 'The beer was getting warm, so we drank it.'

Milton Drew, in Fannie Bay, came through unscathed, but his house was wrecked. Underneath the house, he found that his wife's Volkswagen car now had a bonnet adornment of a fridge that was 'spread-eagled across the bonnet'. He removed it, and months later found that the fridge still worked. But there was tragedy behind his house. He learned that a woman had run out of her house during the storm to collect her pet dog and had been killed by flying glass.

Because he had no family to look after, Milton drove out to his workplace at Winnellie and arrived, miraculously without having a puncture, to find that the CIG plant had been smashed. The yard was covered in gas and oxygen bottles. He stayed there to prevent looting and tidied the yard. He recalled how he felt after the cyclone had passed:

> It was absolute relief. I badly needed a cuppa and a lie down—and a whisky. The next night, people I didn't know blew in and then left again. I'm not a drinker, but on that first night, myself and another man drank a magnum of champagne, a bottle of Club port and a bottle of Kentucky whisky—and I couldn't get drunk!

Driving past the Defence Department aerial farm at Coonawarra, Milton thought that it looked like 'a gigantic mess—like so many huge tangles in a fishing line'.

In Fannie Bay, Elspeth Harvey and her daughter and husband had sheltered in their car. Their choice of refuge, they discovered later when they went into the wreckage of their house, had been guided by luck—a floor joist had invaded the small bathroom where they had considered taking shelter. When they climbed out of the wreckage around their battered car, they found absolute chaos. Being the well-organised person that she is, Elspeth immediately found her insurance papers in their strongbox. She was sitting on the remains of her house steps when friends from Ross Smith Avenue called around. Elspeth recalled:

A caravan had also arrived from two streets away, but our small boat was still in the backyard. Our neighbours' bed was hanging out of their wrecked bedroom. I must admit I did a bit of swearing and I wasn't very lady-like. I tried to maintain a calm facade. The cockatoo was sitting on the barbecue preening his feathers. We found the missing cat.

Elspeth's missing cat deserves a special mention: it was found in their neighbours' letterbox, crying and wrapped, like a feline sausage roll, in iron. It was upside down in the letterbox and couldn't possibly have climbed in on its own; it had been blown into the tiny space. Elspeth added, 'He was a very paranoid cat after that.'

There was a lot of nervous laughter around the place that morning. Everyone decided that a beer would calm them down, and so, in true Darwin style, people gathered and shared a quiet Christmas drink.

Barbara and Greg Huddy, in Stuart Park, had been chased from cover to cover during the storm. They recalled their first impressions after it had passed. Greg said they 'felt numb. We had an overwhelming sense of relief that we had survived.' Barbara added, 'I thanked Christ we were alive.' Barbara recalled that she knew she was sinking into shock, and the pain from the deep cut on her foot was severe. Their one-year-old son Peter's head was full of glass fragments. The Huddys pulled out their camping gas burner and Barbara, who was a trained nurse, began treating people for shock by making cups of tea. Greg recalled, 'A hippie dropped in and began playing Christmas carols on his guitar for the kids, but everyone cried . . . so he stopped.' Their car was intact, except that a wooden pallet had wedged under the rear car springs and lifted the car off the ground. Greg, a customs officer, decided to move his family into Customs House in town, as the building was fairly habitable.

While Greg was off getting Customs House ready that morning, Barbara was a temporary guest in the police station until mid-afternoon. It was a terrible experience, she recalled.

 The Fannie Bay prisoners were also there [in the cells of the Law Courts] and they were behaving like animals. All we had were blankets from the

Darwin Prisons. There was also a woman with no clothes on and covered in gravel rash all over her body where she had been dragged behind a car.

The customs officers were faced with the problem of determining who was still around and who was gone after the cyclone. Some of the people gathered at Customs House seemed to think someone would come and get them. But there *wasn't* anyone. No one seemed to comprehend the extent of the storm damage. When someone said, 'They've got to do something,' Greg Huddy replied, 'No. The "they" is *us*. There's nothing left to help anyone with.'

The Huddys spent several nights in Customs House before Barbara was evacuated. She still recalls the policeman who shared their accommodation. She said that late the next afternoon, after he had spent the day searching the port and harbour for bodies and pulling them from the water, 'he sat down and cried'.

Marshall and Cherry Perron had spent the night in their built-in wardrobe. Marshall was very tired from straining for over five hours with a coat hanger to keep the doors from blowing open. He recalled, 'At dawn we looked out, and I said to Cherry, "Thousands of people have got to be dead out there."' With only three-quarters of their own house left, it wasn't an outrageous statement. Even today, many of the survivors have difficulty in understanding why they were spared. The Perron residence resembled the inside of a blender that had just chopped up half a tonne of glass. Even their bath was full of glass fragments and shards. Cherry thought that their suburb 'looked like an atom bomb had gone off. It was unnatural, like being on the moon.' All of their furniture was gone, and they never found the roof of their house.

Weatherman Geoff Crane came down on to street level with his family at 6.30 a.m. They had trouble getting out, as the wind was still pushing against the fire exit door. With help, he removed a large tree branch that had fallen over his car. The car had been sandblasted down to the bare metal. Another staff member's car had been pushed sideways against the gutter and flipped on to its roof. By mid-morning, the Cranes were able to make reasonable progress back to Rapid Creek, as a bulldozer had cleared a path along Bagot Road. At one stage Geoff thought he was on the wrong road, as there were simply no reference points left to guide them. He recalled, 'I felt I was getting shell-shocked as we got closer. The kids were quiet . . . I think we were all simply over-awed.'

Geoff's decision to bring his family into town on Christmas Eve proved very wise, as their house was totally demolished. Barbara said, 'The floors were swept clean.' Geoff added, 'We probably would have died had we stayed there.' All that was left of their house was the cyclone bolts and, literally, the kitchen sink. Everything was in the backyard. They quickly decided they couldn't stay underneath their wrecked house, as many others were able to do, so they headed off to the Rapid Creek Primary School just around the corner, broke in and set up a refuge there for survivors. Several other meteorologists lived nearby and they all gathered at the school, got things organised and even had a Christmas dinner that night.

Way out at Humpty Doo, Jenny Lonergan and her husband walked over a kilometre back down the road to their wrecked house from their car that had been parked between two road trains. Some cattle on one truck had died during the night and the truckies reckoned it was because they couldn't breathe in the wind. Jenny recalled that, in the morning stillness, she felt 'lucky to be alive. The house was just a shambles. In the middle of the road was our TV. All John's books were everywhere. I didn't cry—I was just stunned.'

Everywhere in Darwin, people were consoling each other and sharing their meagre possessions. Clothes were given to total strangers. It wasn't a time to whinge or moan, as everyone was in the same boat. It was a time to work together, help each other, and be compassionate and sharing. And that's what they did. When the nation heard what had happened, the rest of Australia did the same.

12 HAM, JAM, LAMB OR SPAM?

TWENTY-FOUR HOURS after Tropical Cyclone *Tracy* had passed, the disaster relief plan was under way, not only in the smashed and shattered town of Darwin, but right across the nation. Once the head of the newly formed National Disasters Organisation, Major General Alan Stretton, arrived and was quickly appraised of the situation, a huge relief task force was deployed and came to Darwin's aid.

The problem with the storm damage in Darwin was that it was so extensive; there wasn't much left to do anything with. The roads were choked, and water and power were almost non-existent, save for some diesel- and petrol-powered generators. There was precious little habitable accommodation and about half of the population was homeless. Medical supplies and facilities that hadn't been blown away were exhausted, and there was nowhere to store salvage. Almost everyone in Darwin was affected to some extent by the storm, and so relief and assistance had to be external and self-sufficient. The destruction was so total that outside assistance was the only way to begin recovery. The ideal force in that regard was the Royal Australian Navy, which could provide all manner of support from offshore and wasn't reliant on the town for water, food, power or accommodation. So, the RAN was first in after only a few days and busily prepared the way for the military and civilian relief task forces that would arrive in early January.

The death toll from Cyclone *Tracy* was officially fifty, with another sixteen lives lost at sea. Although other cyclones in the Australian area have caused greater loss of life, mainly at sea among the pearling and fishing fleets of earlier years, none has caused damage of the magnitude suffered by Darwin. Approximately 20,000 people were rendered homeless when 90 per cent of the town was flattened. The damage bill was estimated at between A$500 and A$600 million dollars in 1974 terms—an equivalent of between A$2.6 and A$3.2 billion in 2001.

Meanwhile, it was up to the survivors themselves to get along as best they could until that assistance arrived. To get urgently needed shipping in and tied up at the wharves, harbour master Carl Allridge had to clear under berths with soundings. Every wharf was damaged to some degree, and sunken vessels impeded the wharf approaches. The wharf storage sheds had lost their doors. On the Iron Ore Wharf, the pipelines carrying water and electricity were damaged. A Bailey bridge was eventually erected to allow access to the Iron Ore Wharf, which had a huge gap in the approach. The liner SS *Patris* would soon make this wharf its home during the clean-up and reconstruction phase. The immediate task was to search for survivors, as at least twenty people were initially reported missing at sea. Several vessels provided invaluable assistance in searching for survivors and conducting soundings. In his report of 4 January 1975, Carl Allridge wrote:

> . . . divers are unable to distinguish one vessel from another in the heap of dangerous jagged rubble remaining in one particular area. At least five vessels went through the approach to the Iron Ore Wharf tearing away over 50 metres of decking and piping. These vessels have not been found. At least one vessel went under the approach to the Fort Hill Wharf.

On Boxing Day, not knowing whether the wharves were clear, it was arranged that the *Miss Rankin*, a survey vessel, would carry out soundings. She also assisted in other work, such as transporting people from Mandorah, on Cox Peninsula, and searching for survivors. The *Lady Cynthia*, a rig tender that arrived on Christmas Day, also did invaluable work in surveying, towing, searching and other tasks, including rescuing two men from the upturned *Frigate Bird*. The *Northern Tide*, a rig tender that arrived on Boxing Day, also helped in searching the port. The *Darwin Trader* also gave assistance when she arrived on 27 December by clearing the wharf and sheds, feeding the port workforce, accommodating people on board, and taking over the communication centre when the *Nyandra* sailed. The *Mutual Enterprise* also assisted, as did the prawning vessels—such as those that supplied power where possible. These include the N.R. *Junction*, N.R. *Mitchell* and the *Raima Kathleen*. Following the tropical cyclone, the *Eiko Maru*, a prawn-carrying vessel, berthed and took the body of Koji Yoshada, operations manager of Northern Research, who was lost on the N.R. *Kendall*, back to Japan. A fishing vessel, *Kempira Maru 15*, had been driven ashore on Bathurst Island. The crew of eighteen survived and ended up at Bathurst Island Mission after being rescued by the *Raima Kathleen*.

Most port workers arrived for work on Boxing Day and set about restoring operations. The *Nyandra* provided a midday meal and the port was soon functioning again. The immediate tasks were to ensure that the approaches were cleared of debris, pipelines for handling fresh water and fuel were replaced, the pilot service was maintained and the wharf shored to accommodate heavy cargo. The huge relief effort that was about to descend on Darwin would need the port and wharf facilities to ensure a constant supply of materials and food during the wet season, when the main highway was often cut for weeks at a time.

In town, a long queue of people formed outside the police station in the city. They were redirected to schools in their residential area and told to sign up for relief and work party duties. John Woodcock recalled that the police 'liberated' food from wrecked supermarkets and stores for the relief centres. Sanitation was a high priority, and many places had work parties whose sole responsibility was providing seawater with which to flush toilets. Public tap facilities were established along the main water pipeline that paralleled the Stuart Highway. These had to be controlled to avoid disputes and to establish some order. But in general, people were compliant and cooperative, with only a few exceptions brought about by stress, fatigue and frustration.

Police forces from other states and the Australian Federal Police soon supplemented the Northern Territory Police. The NT Police were a small, well-known group who understood and tolerated the somewhat different behaviour and style of Territorians. As John Woodcock recalled, 'The visiting cops were wearing sidearms and were intimidating.' The NT Police wore firearms for only a short period when roaming packs of dogs were a cause for concern and had to be destroyed. The police ran the school relief centres on behalf of the NT Administration and were a visible sign of authority amid the organised chaos. Everyone spoke highly of the NT Police and their efforts during this extremely difficult period.

There were isolated incidents of looting during the post-cyclone period. Sergeant John Woodcock thought that of the 10,000 people who remained in Darwin working on the clean-up, there were only about twelve arrests. John remembered one man was sentenced to two months' imprisonment for stealing alcohol. The police took a circumspect view of what was considered looting. Anything that could be construed as providing clothing or shelter was considered more a case of 'survival' than of 'looting'. However, Eddie Quong had a run-in with some unpleasant types who were helping themselves to his possessions and he was about to use a firearm on the two men when his eldest son stepped in and prevented what could have been a nasty incident. The villains left empty-handed.

Several tyre companies provided round-the-clock support to the essential services to keep their vehicles on the road. To venture out on a trip of more than half a day meant taking at least two spare tyres, as punctures were guaranteed. The tyre companies had men working non-stop changing wheels and tyres, and providing a working stock of spares. Young Peter McIver worked for two months at a service station after the cyclone, just changing tyres. Soup kitchens were quickly established to provide food for anyone and everyone. Most kitchens were at the local schools, but several were also set up in town where there was a large working contingent during the day. Karen Jurek worked in a soup kitchen when she returned to Darwin after spending only ten days in Dampier in Western Australia.

The power station in Darwin was a huge affair that provided the town electricity through a bank of five enormous diesel generators and turbines. The power plant had been put out of action and had to be refurbished. Vast amounts of saltwater had penetrated the cooling system when the saltwater desalination cleaning system failed during the storm. Jenny Lonergan's husband John worked on the turbines for eight weeks to bring them back to operational condition. Jenny stayed at the power station working as a cook for the men who were trying to get power restored as quickly as possible. It took several months before the power could be safely reticulated, owing to the massive damage to the power grid. Jenny recalled that one dark, quiet night just a few days after the cyclone, she was 'sitting upstairs in the power house crib room, and from a distance I heard a man crying his heart out'.

Driving out to the suburbs at night was like entering a ghost town. There were very few lights, usually hurricane lamps or gas lamps burning in isolated dwellings. The chatter of small petrol-powered generators would last until around 8.00 or 9.00 p.m., when most people would call it quits for the day. People slept wherever they could stay dry. Greg Huddy's family slept atop desks in Customs House. Julie Tammo and her husband slept on a billiard table in the fire station. Peter and Ann Gray made a 'bad move' when they slept on the fibreglass insulation material in the Wanguri Telephone Exchange building and itched for days. John and Sue Ryan were bunked down at the Alawa School for three days and nights, waiting for Sue and baby Daniel to be evacuated. They recalled that the worst part of this enforced community living was the constant flow of rumours that tended to be unsettling and which upset many people who were already traumatised. Medical teams began to visit the relief centres, giving compulsory injections for typhoid, cholera and tetanus. They also tended minor injuries—still without anaesthetic. The fire brigade was delivering potable water to relief centres, using their pumper trucks to carry the water.

Many people had no clothes apart from what they were wearing on the night of the storm. Rowan Charrington went to Casuarina shopping centre to get some underwear and clothes for Marcia, who was looking after their young son. He explained to the policeman standing guard outside the store what he wanted, 'while trying not to look like a deviate', he said. He was permitted to take what he needed and returned with a size-12 dress for Marcia. Unfortunately, it was a child's size 12!

Health was a primary concern for the authorities. The combination of the wet season, a lack of sanitation and the rubbish strewn everywhere posed a significant threat for outbreaks of highly contagious and potentially fatal diseases such as cholera and typhoid. The authorities decided to evacuate the town of non-essential and sick and injured people. Anyone who didn't have a job helping to restore the town, who was unable to work, or who didn't have shelter, was to be evacuated to one of the Commonwealth 'refugee centres' that had been established in defence establishments around the country. The sick and injured would be evacuated first, then pregnant women and those with young babies, then women with young children, followed by 'non-essential' people.

The recently elected Northern Territory Legislative Assembly, which Marshall Perron described as a 'debating society' (it had no real powers, and only provided advice and community feedback to the administrators), met to come to grips with the task of getting Darwin back on its feet. Grant Tambling recalled that they soon began working twenty hours a day to try and keep on top of the myriad of tasks that required attention. 'Without the Commonwealth support it would have been impossible, because the whole community was affected,' he said. A reconstruction commission headed by former Brisbane lord mayor Clem Jones was formed to rebuild the town. There were rumours that Darwin would be razed and rebuilt, so extensive was the damage, but the fact that the basic infrastructure facilities such as the port, airfield and roads still existed helped to keep heads cool in that regard.

The first plane in on Christmas night was a Boeing 707 carrying the head of the NDO and officials representing the Commonwealth government. Hardly a plane left Darwin in the next seven days that wasn't crammed with evacuees almost to bursting point. All manner of aircraft were used, from RAAF C130 Hercules aircraft to 707 and 747 commercial jet aircraft, to evacuate the citizens of Darwin. The New Zealand and United States military also provided aircraft. Because passengers came on board with only hand luggage, comprising a change of underwear and some very basic needs, the passenger-carrying capacity was extended. Unfortunately, this was not by adding seats, but bums to seats, and for the first few days almost everyone who was flown out had a baby or young child on their lap. On Friday, 27 December, a Qantas Boeing 707 lifted out three hundred and twenty-seven people on an aircraft that would normally carry one hundred and seventy-three. A day later, a Boeing 747 jumbo jet that routinely carried three hundred and thirty-three passengers took out a record six hundred and seventy-three, only to see the pilot of another 747 take an additional person, and claim a record six hundred and seventy-four souls on board, on Sunday, 29 December. Greta Quong and three of her children went out with only one carry-bag between them. Eddie Quong lasted about another three weeks before he fell ill and was evacuated to Adelaide with shingles and spent two weeks in hospital.

Dianne Ferguson lived on the RAAF base and, because people were close by and easily transported, they were soon on their way. She was given a seat with her two young children on a medical evacuation (medevac) Hercules. The medevac aircraft came with stretchers, and a medical team of usually one doctor, several nurses and a few medics. She recalled:

 It was pretty horrific. The plane was full of stretchers and about only 10 per cent of the people on board were walking wounded. The rest of the passengers had drips . . . There was one lady without a leg. It was an awful trip. I flew to Brisbane; the Red Cross was waiting. Then we had a short trip across to Amberley and I stayed the night in the airmen's quarters.

Dianne's family lived in Grafton, and they came up and collected her and the children. She stayed for six weeks and then went back to join her husband in Darwin once they had a place to live in that had a roof.

Barbara Huddy's occupation as a midwife got her on a Fokker Friendship on Boxing Day that was a 'maternity special'. Almost every woman on the plane was pregnant or in a post-natal state. Barbara had heard the stewardesses saying they needed a midwife, and she volunteered. The plane had little fuel and couldn't refuel in

Darwin. The aircraft had to travel to Brisbane via Katherine and Mt Isa. She flew nursing three children. She recalled:

 The mothers were half-crying, everyone had sore bums, no milk, and there were also a few injured people. Our daughter Suzy was nursed by one of the navy patrol boat captains all the way to Brisbane. I was kept busy looking after the women, of whom two were in labour, but luckily they didn't give birth.

What the people waiting to receive the survivors didn't fully appreciate at the time was that they were refugees in every sense of the word. They were homeless, they had come from a place that had been totally demolished, and they had been scared to death. Many were still in shock and suffering both physically and mentally from their ordeal. It took many survivors days before they could comprehend the enormity of what they had been through. When Barbara finally arrived in Sydney and was driving down to Canberra, she noticed that her daughter's legs had been 'ring-barked' by generously donated socks that were probably a couple of sizes too small for her legs. They left a permanent legacy of the cyclone.

Josephine Foreman was evacuated on 27 December by an RAAF Hercules. She recalled that the plane was 'ankle deep in sick . . . and cold. I realised when we arrived in Melbourne that Australia was stunned.' After flying on to Richmond airbase outside Sydney, the women were placed on an RAAF bus. Josephine recalled that it was disconcerting, because:

. . . everyone was telling us what to do but not explaining why we were doing it. A policeman climbed aboard and asked us for messages to send back to Darwin. No one spoke for a minute, and then some started to cry and before long he had a busload of women in tears.

On the trip from Richmond to Newcastle by car, Josephine's relatives couldn't understand that she just wanted some peace and quiet and didn't want to talk about her ordeal. On arriving at the family's home, when she saw her father waiting, she let out her emotions for the first time in three days and collapsed sobbing into his arms. She slept for twelve hours.

Some women disappeared interstate while their husbands were on work parties. John Ryan returned to the shelter at Wagaman to find that Sue had left with their son Daniel. She recalled, 'I walked out to the plane and I had no idea where it was going . . . but I knew it was out! I was desperate to get out with Daniel.' She arrived in Mt Isa at 2.00 a.m. and recalls the 'soup kitchen, and the wonderful first meal—my first in four days. It was curry and rice.' Daniel was whisked out of her arms and taken for a wash, and Sue was given a bundle of clothes and nappies.

Chris Kingston-Lee had been away on 'queue duty', lining up for food for his wife and young son. When he returned, he found they had gone. For two days, he didn't know where they were, though he knew they had been evacuated. They ended up in Sydney, where they were housed in the quarantine station at North Head because the authorities feared a cholera outbreak among the refugees.

The RAAF Hercules is a pressurised aircraft but has very little insulation against noise. Military passengers normally wear ear protection. It is almost impossible to converse in the cargo compartment without shouting in the listener's ear. Teenager Mathew Coffey and his brother had suffered ear damage during the cyclone, and during the days immediately after it they had blood in their ears. On the flight south in the propeller-driven Hercules, they jammed Stimorol chewing gum into their ears to ease the pain and reduce the noise of the high-pitched whine of the turbines.

Janice Woodcock was in another early group that went out from the NT Police families. The rationale was that the policemen would be working long hours and needed to be on duty around the clock. Janice recalled that the plane had 'only women and children on board and twice the normal capacity, with everyone two to a seat'.

Some people not slated for early evacuation went to extreme measures to board a plane. A number of men tried to board one aircraft by posing as injured evacuees or women, but were discovered by the authorities and dealt with harshly.

Lesley Mance was evacuated by commercial passenger jet aircraft to Sydney. The Salvation Army gave her a dress to wear that was more suited to an elderly woman. The fashion-conscious teenager was mortified! Her injured sister-in-law stayed in Darwin Hospital for a week. Lesley recalled that, on the plane, she nursed a baby with one blue and one green eye. 'People were sitting on the floor of the plane. Mum was evacuated to Brisbane Hospital and was in there for three months with her cut leg. No one talks about it today at home.'

Peter and Ann Gray were evacuated as wounded survivors. A photograph of them arriving in Sydney in wheelchairs appeared on the front page of most daily newspapers. Peter's foot injuries had steadily worsened. He had been partially crucified under the roof of his house. He had struggled about for a while, but collapsed at Casuarina High School and was unable to walk. He also had a bad reaction to the penicillin injection and was stretchered out with his wife on a medevac Hercules. Peter recalled, 'The Herc was full to capacity. Most people sat in the web seats, and I remember there was even a humidicrib on board.' When they arrived at Concord Repatriation Hospital, they were somewhat embarrassed to have about fourteen doctors waiting on just three patients.

Marcia Charrington was evacuated, but had a difficult time just getting on the plane:

> I went out on a jet, via Darwin Primary School. There were awful delays, standing in the hot sun, and then we spent four hours sitting in a bus . . . The Salvos were terrific, handing out fruit juices, but it was a horrendous wait. There were five planes sitting on the tarmac and they were going all over Australia. We were going to Adelaide. I was so nervous about getting on the right plane. The plane had no water or food on board. There was a Greek lady in the next seat who took up a lot of room, and on the other side of me a woman was unconscious.

Marcia arrived in Adelaide at 4.30 a.m. and was met by a host of people ready to hand out clothing, food, money and basic essentials for sanitation and personal hygiene. She was feeling 'off-colour' during this ordeal and went for a check-up, only to find that she was pregnant.

The injured Clarkes, who had been blown out of their house with their children in their arms, were evacuated all the way back to Gisborne in New Zealand free of charge. Gisborne was where

Graeme's parents lived and could best provide a temporary home. Because Graeme worked for TAA he knew exactly how many people could fit on a 727 jet plane; he must have been showing some concern, as the plane was 'packed to the ceiling' with people. When a member of the cabin crew recognised Graeme and saw the look of concern on his face, he reminded him about the lack of luggage. The Clarkes flew into Sydney at 1.00 a.m. and were given overnight accommodation at the Callan Park psychiatric hospital. Graeme recalled:

> It was chaos; the staff wanted to isolate people and split them up. They didn't realise the trauma we had been through . . . Anyway, I stood my digs and kept the family together . . . Getting to the airport to go to New Zealand was another drama. We couldn't get a taxi, and eventually someone sent a stretch limousine to take us to the international airport.

On arrival at Mascot airport, they were given a sum of $150 in aid. They travelled First Class to New Zealand, dressed in 'fairly shabby gear'. Graeme was resplendent in shorts that were far too big for him and two different-coloured thongs. He fondly recalled the wonderful treatment they received from everyone who discovered that they were *Tracy* survivors. But once in New Zealand, Graeme's injuries became septic. Tricia, who had a seriously gashed right arm, was given a cholera injection in her left arm and soon found she had no arms capable of working!

Trawlerman Rob Perkins left Darwin in style in the company Lear jet. The prawn trawler fleet was defunct and there was no work for him in town.

Laurie and Shirley Gwynne went out to Berrimah, to the forestry research station, and made themselves comfortable where, even if the conditions were basic, it was at least dry and they had hot water for showers. Shirley was very sore from her escapade in their backyard. She lined up for two days at Wagaman School, trying to get on a plane. She remembers vividly the look on a young army medic's face when he had to give her a tetanus injection and was worried that he would hurt her. The doctors and nurses dispensing medicine asked Shirley if she wanted the contraceptive pill. She recalled with a wry smile, 'After all the battering and bruising I had taken, sex wasn't high on the agenda.'

Shirley's evacuation had some different twists. She recalled wearing paper panties and a halter dress that was covered in carbon paper smears where their neighbour's office supplies had landed among their scattered clothing. When they finally got to the airport, she was 'thrown off two planes because they were overloaded' and finally got on to a plane headed for Brisbane. While standing out on the hot tarmac, she was wilting badly. A passing policeman saw her predicament and took her aside. He got some cold lemonade and a meat pie and buns for her and the children, but Shirley ate very little, as her hungry kids devoured the lot. It was three days since she'd eaten anything. The policeman then escorted them straight on to the plane and sat them down. She recalled that they sat in the plane for a long time while waiting for Prime Minister Gough Whitlam's plane to land. 'He didn't win too many votes that day,' she said. During the plane trip the kids shared an apple, the only food that was provided.

On their arrival at Brisbane at 3.00 a.m., the Salvation Army women tried to take the children from Shirley to give them a wash and a change of clothes. 'I lost it then,' Shirley recalled. 'I was worried that I would lose them, because we had no identification and there were people everywhere.' The children were quickly reunited with the distressed Shirley, who tied them to her wrist. The group then spent time with relatives on the Darling Downs, west of Brisbane.

Diane Kerntke had a smooth passage to Adelaide on Boxing Day in a plane designated for army families. Her dramas began when the army bus that was dropping people off around Adelaide took forever to complete its task. It was very late at night, everyone was exhausted, the bus trip had already taken three hours and they still had more stops to make. In frustration, Diane and another woman and their children got off the bus and managed to get a cab. Not knowing exactly where she was, it took Diane some time to get her bearings. When she arrived at her destination, the cab driver refused to take the $16 cab fare when he learned they were from Darwin.

Some of the people who assisted those in need were true Samaritans. Karen Jurek recounted the story of a woman by the name of Joy (last name unknown) whom she met when she was evacuated to Port Hedland. Karen needed to get from Port Hedland to Dampier, a distance of 235 kilometres, by road. When she arrived in Port Hedland, the woman volunteered to drive her all the way to Dampier on what was then an unsealed road. When Karen asked her why she would go to all that trouble, and for no payment, the woman replied that it was her contribution to the cyclone victims. She gave Karen clothes and some money. When they arrived in Dampier, she refused the offer of hospitality. Karen recalled, 'She left to return to Port Hedland to meet another flight.'

Judy Pickering and her son were put on a medevac Hercules to Adelaide, and then on a commercial jet to Melbourne. Her son Ben had contracted conjunctivitis from the wet dirt that blew down through the floorboards in their shelter. When the doctor noticed his swollen tummy, Judy had to explain that all he had eaten for the last day or so was Savoury Shapes biscuits and oranges.

Milton Drew was kept busy in Darwin distributing LPG appliances and gas for cooking. He flew back to Brisbane on one of the huge American Starlifter cargo planes with only four people on board. The Starlifter had been shuttling stores and heavy equipment into Darwin. He recalled, 'It was an ex-Viet Nam plane and looked like it was being driven by teenagers.'

Julie Tammo recalled that when she boarded her evacuation flight, she was given a yellow 'nuclear attack victim' card issued by the Department of Defence for the purposes of registration and identification. Looking down on the remains of the shattered town as she flew out, she thought it seemed very apt. The 'seriously packed' commercial jet airliner that she flew out in 'nearly went through the runway when we landed in Brisbane. Every seat had two people in it.' She eventually got to Adelaide but couldn't handle staying with her in-laws, where two other women and eight children were already being billeted. Heikki came down on R&R leave, and they bought a caravan and towed it back to Darwin. The Tammos had been planning to go on a trip to Europe after Christmas, but they bought the caravan instead.

By year's end, some 20,000 people had been airlifted out in a period of six days. By midnight on New Year's Eve, Darwin airport had managed to dispatch one hundred and forty-five flights out of town. Of these, eighty-nine were scheduled flights, sixty-four were military aircraft and twelve were other non-military flights. It was a fantastic effort that relieved the pressure on the strained facilities and huge workload off those who remained behind.

Another 6000 people evacuated Darwin by road. Some people didn't even wait to be helped out of town; they simply threw whatever they had left in their car and took off—many never to return. The government provided considerable aid and support to those leaving by road, including the provision of four brand-new tyres to avoid having people travelling long distances on tyres that might have been punctured and repaired. A roadblock was established at Katherine, and at several other entry points on the major roads, to prevent people from driving into Darwin without special permits.

Chris Kingston-Lee had stayed in Darwin until his family was evacuated by air; he then drove down The Track to Adelaide. He had lost more than his house—his motorcycle business was in ruins. He was 'not a happy camper', he said. On the road well south of Darwin, his trailer blew a tyre and rolled over. He was so upset at this 'final straw', he simply cut the trailer loose and left the remnants of his goods and chattels in the dry desert sand.

Judy Watson went into delayed shock after three or four days. Her children were evacuated on New Year's Eve to be with her parents in Beenleigh in Queensland, and Judy and her husband decided to drive to Brisbane via Townsville in a classy red Holden Torana XU-1 sports coupé to sell for a friend. Judy recalled:

 It was more traumatic than the cyclone! We went via Katherine on 3 January, stayed overnight in the showgrounds, and when we eventually got to Mt Isa we were flooded in! We then caught a Hercules to Lavarack Barracks in Townsville to wait for our cars, which would come on the train. That took another three weeks. We finally got into Brisbane in early

February. At Mt Isa, the Salvation Army ladies were all lined up and waiting to give us clothes and food, and so on. One lady offered me sandwiches. I'll never forget it. She asked me, 'What will it be, dear—ham, jam, lamb or Spam?'

Laurie Gwynne stayed in Darwin, hobbling around and organising the clean-up with his forestry work teams. But his broken foot soon became worse and he decided to drive out. He was given needles to self-inject to stop any infection. He took their brand-new Mazda car and travelled in a small convoy that was heading towards Townsville and further south. He was given dole money and fuel vouchers. Like many others, he got to Mt Isa and was flooded in. He slept on the train platform, was woken up at 'some ungodly hour', loaded his car on to a flat-top wagon and went on the same train to Townsville. He had to get his car serviced in Townsville because it was brand-new, and an obliging and sympathetic mechanic took him to lunch when he found out who he was. Laurie then drove down to Brisbane, but on the way he stopped in at Mackay to visit a mate. He ceremoniously handed over the front-door key to his friend's house in Darwin, which they had been minding, but informed him that there was no door—or house—to go with it.

Greg Huddy left Darwin on New Year's Day, also via Mt Isa—'and several pubs'—in a convoy with 'four other blokes'. It took him ten days to get to Brisbane, 'travelling at 80 kilometres per hour' in his car that was crammed with the only belongings they could find in their backyard. He also took the train to Townsville from Mt Isa. He was given new tyres for his car in Townsville and recalled being overwhelmed by the generous assistance. Greg's favourite story is of when he got to the roadblock at Bachelor, south of Darwin and down The Track. A large policeman with a shotgun was manning the roadblock and waved their convoy to a halt.

 This copper, wearing stubby shorts, a shotgun and a hat, insisted we had a beer and a meal in town. We said we'd like to get going. He scowled and said that we were going nowhere, because there was a bunch of ladies in town who had been up all night preparing food and he wasn't going to see them disappointed . . . So we went into town.

The majority of people left in Darwin were men. Everyone was working—and working bloody hard. Greg Huddy recalled an amusing incident when a man who wanted to renew his licence to operate a front-end loader called in at the Customs House looking for the police. He had the wrong place, but he was so keen to get out and on the job that Greg stamped and signed his old licence with a Customs Port stamp—and away the fellow went on his front-end loader.

Kevin Jurek worked for the Northern Territory Administration as a foreman, and his work team of electrical workers was flat out. Kevin recounted, 'There was a lot of freight coming in to rebuild the supply system. I spent my days organising work teams to set up generators for stores and essential services.' He organised a big army diesel generator that weighed almost 2 tonnes that ran half a dozen houses in his street in Jingili. When Karen returned to Darwin from Dampier, she said she would often go out 'scrounging'. She 'borrowed' louvres and curtains from wrecked houses in the northern suburbs.

Chris Collins was working on delivery trucks going to the relief centres. About three days after the storm, he was at Casuarina High School.

 We were buggered . . . We had been working long hours every day and it was very hot. There were some blokes there sitting on their backsides and they refused our request to help unload the truck. A NSW copper asked what the problem was. We told him. He cocked his gun . . . then they worked!

Chris worked out at the Berrimah research farm, where the group slept on desks—but they had solar hot water. They offered the policeman who had helped them a hot shower, which he gratefully accepted. That night, after a meal of baked beans and red wine, the policeman was telling stories about trying to catch looters and so they provided him with bird shot which they used to scare the birds of the experimental crops at the farm. The policeman reported back that it was hugely successful in deterring would-be criminals. But one of Chris' enduring memories is of driving down the highway and seeing nude people out in the open showering under a water main.

Tony Pickering and his mate Peter Coombe were sitting around one of the schools near Ludmilla when a colleague by the name of Gough Letts asked if they would volunteer as bus drivers. They eagerly jumped at the chance, only to find that they were driving a latrine truck that consisted of dozens of open pans that needed to be taken to the waste disposal area. Tony was waiting outside a ladies' public toilet to collect the pans when a passerby threw an empty bottle into the back of the truck, thinking it was a rubbish truck. The ensuing splash covered Tony in a most undesirable fragrance.

Elspeth Harvey stayed in Darwin, as she was needed to help get the schools started again. They reopened for limited numbers of children in late January. Elspeth spent her days prior to that cleaning library books and re-cataloguing the stock that had survived the storm. She was also on a team of people that was preparing the pre-schools. She recalled feeling that she never wanted to clean another room. Elspeth and her husband stayed in Darwin and lived in the remains of their house, which she said made cooking in a rainstorm a real challenge. Her husband built a cage to house their pet dogs, to stop them from wandering and being shot as strays.

Carl Allridge's daughters were evacuated, but his wife, who was heavily involved in the Red Cross, stayed behind and was appointed to the Darwin Reconstruction Commission. She was also awarded one of the Red Cross Society's meritorious awards for her untiring and dedicated work after *Tracy*.

At Rapid Creek, where devastation of the houses was almost total, the local school was packed with survivors. Barbara Crane felt a keen loss when she learned that her tennis partner, Jenny Wood, had been killed. Among their group they soon organised people to go on 'queue duties'. It was necessary to queue for water, food, clothing and even inoculations. People would go and queue, and then relieve others after an hour or so of standing in the hot sun. Everyone soon had duties; some, like Marcia Charrington, drew the short straw by being on cooking duties between 2.00 and 4.00 a.m., which included keeping fires burning to boil water and prepare food. Geoff Crane recalled that some of the mets who stayed in Darwin went for their cholera injections and then walked up eight flights of stairs to their office and promptly collapsed. He had found his television set in the yard, but it didn't work when he tried it. He thought he would dry it out, and took it into the office where it sat for months. He came into work one day and the shift was watching a sporting event on TV, and he noticed it was his TV. Years later, in Tasmania, it stopped working and a bemused repairman asked Geoff how so much sand had ended up in the back of the set.

Nick Paspaley eventually got back to his boat, the *Paspaley Pearl*, in Sadgroves Creek when the roads were cleared. It had been reported that there were bits and pieces of fibreglass boat up on the Frances Bay shoreline and he thought his boat had been wrecked. He ran into a long-lost cousin and was able to borrow a dinghy. He recalled:

> I drove around the waterfront as much as I could, and there were no boats. Every boat that I saw was up on the rocks wrecked or sunk in the harbour . . . and I went up the creek where I anchored it and there it was still in the creek. It was about 200 metres in the mangroves. It had dragged the anchors . . . and dug a channel backwards through the mangroves . . . and was sitting upright. Its anchor chains were right over the top of the mangrove trees where they had gone at high tide. So we swam down, we cut the trees out and winched in the chains. We cut the trees off, and on the second tide we were back floating in the middle of the bay . . . All the paint was sandblasted off the front and the deck rails, but the pepper and salt shakers were still sitting up on the table! I couldn't believe it! I started the engine, had a hot shower, had a sleep in the bunk, opened up some baked beans and cooked some baked beans and toast. The galley was still perfect. It was the only decent meal I had had for days. It was unbelievable—the boat was just unscathed. I hosed off the deck with the deck hose and it was a new boat—beautiful.

It was a small victory in a battle with nature. A lot of Nick's mates had been drowned and their boats lost at sea. He had tried to talk some of them into leaving their vessels, but they had stayed on board and lost their lives. Five seamen were lost on the *Booya*, two on the *Frigate Bird*, three on the *Flood Bird*, two on the *Mandorah Queen*, and one each on the *Darwin Princess* and the *Charles Todd*. Several others were still missing.

Darwin was on the road to recovery, but it would be a very long time before it would be back on its feet again. The devastation that Cyclone *Tracy* caused in one night would take many years to repair.

13 EPILOGUE

DARWIN WAS CHANGED forever by Tropical Cyclone *Tracy*. Gone was the ramshackle, jerry-built country town. The population changed when many people left town and didn't return. Some moved elsewhere because they couldn't face the prospect of another storm—and another night of fear such as they had experienced in the cyclone. Others left because they had no job left in town and needed to start afresh where there was a chance of establishing a new livelihood. Yet, many people didn't even consider leaving; Darwin was their 'home' and they weren't going to be chased away. They loved the casual lifestyle and thought that the risks were worth it; besides, how often would a *Tracy* come along?

In the aftermath of the cyclone, rumours abounded of mass graves of hundreds of people who had died in the storm and been buried secretly because the government didn't want people to be too frightened to remain in the north. In October 1977, a sensationalist article entitled 'The Missing 500' appeared in a tabloid newspaper, the *Sunday Observer*. The article quoted a Mrs Carol Ford, who worked as a barmaid at the Territorial Hotel, who said that a man (name unknown) had told her that he was a truck driver on a regular Darwin to Cairns run during the period immediately following the cyclone. He alleged that in early 1975 he took his refrigerated truck and trailer crammed with five hundred bodies wrapped in plastic bags to a secret destination somewhere near Adelaide River. When he got down the track about 160 kilometres from Darwin, he told Ford, the bodies were dumped into a big hole and bulldozed over. He said that he had been sworn to secrecy over this gruesome affair. Ford also claimed to have seen the mass grave sites, allegedly three in all. She told the *Sunday Observer* reporter, 'The ground was level, and new grass was growing through. New trees had been planted at the site.' She said that the truck driver had told her that many of the bodies were children.

While I was in Darwin doing research for this book, I had a telephone call from a Mrs Christine Cox, who said her husband, policeman Harry Cox, was working at Maningrida. She told me that her husband had been involved in an extensive search of the alleged burial sites and they had found nothing. Indeed, if five hundred people had been missing, their next-of-kin and others would have come forward, and the Darwin Coroner's office has no record of missing persons from the cyclone apart from those known to have been lost at sea. In any case, why would there be a cover-up? Some people have suggested that it was to avoid huge compensation pay-outs for the allegedly shoddily-built dwellings that were mostly NT government and Housing Commission houses. But there is no evidence to support these outrageous claims.

The search for bodies continued for months after the cyclone. Carl Allridge recalled that one boat was found some eighteen months after *Tracy*, with 'one-and-a-half bodies in it'. Another rumour that was found to be false involved a social group being lost on the *Mandorah Queen*. The *Booya* was never found. Some sixteen souls were lost at sea, presumed drowned. In Carl Allridge's view, 'The cyclone changed Darwin. Lots of people left. There were many marriage break-ups and lots of new houses.'

One theory about the cyclone that does hold some water is that if it hadn't hit when it did, in darkness, then more people may have lost their lives trying to do things they didn't do that night because it was too dark to see. Several people were killed, and many more injured, when they left their shelter and went outside during the storm. I am convinced that had it not been so dark, then more people may well have 'chanced their arm' and tried to go outside when it was fatal to do so. Rowan Charrington believes that most people 'did what was right for them at the time', whether it was to seek shelter in a car or storeroom or hide in a wardrobe. Every situation was different; no two households experienced exactly the same fate.

Also, Darwin was lucky with *Tracy* because when the tropical cyclone made landfall the resultant storm surge—a rise in sea level due to the very low atmospheric pressure and the stress of the wind on the sea surface—had little impact on the town or its residents. The maximum height of the storm surge measured in the Darwin Harbour was 1.6 metres at about 2.15 a.m. At Fannie Bay, there is evidence that there was a surge of between 1.7 and 2.0 metres. Fortunately, 24 to 25 December was a period of tides and, even though the maximum storm surge was associated with a high tide, the water level did not rise enough to be a problem.

Recounting their experiences of the second wind is still, for many, very traumatic. Some simply cannot face the pain of talking or even reflecting on those petrifying hours. It was a terrifying ordeal for almost everyone who lived through the storm. John Auld thought that Mother Nature had been hard on them. A 'twister' (tornado) in the Northern Hemisphere only lasted minutes as it tore through an area, he commented, but *Tracy* had belted them for four or five hours. His wife Helen lives daily with memories of nearly being killed and says she will never return to Darwin. She regrets the loss of their lifestyle in Darwin, where they had lived for ten years, and the close friends they had made. 'The memories of that terrible night bring back tears,' she said.

Many people's attitude to life changed after they had looked death in the face that night. One man waiting at the airport, who had just been kitted out with clothes by the Salvation Army, was heard to say to his friend that he felt 'like a king'. When he was asked why, he replied: 'Mate, I've lost the lot, absolutely everything. But now I'm wearing $6 underpants and I've never worn anything like them in my life!'

Tricia Clarke has lived in Darwin her whole life. She believes that Darwin is now a better town, created by 'a new wave' of people bringing a different style to the northern city. A very strong community spirit was forged after the storm; everyone helped each other and provided their own form of counselling at critical times, she said. Indeed, Tricia had a friend who she said 'felt left out' because she wasn't in Darwin when *Tracy* struck. Despite her injuries and the fact that it was the most frightening night of her life, Tricia added: 'I wouldn't have missed it for the world.' While she found this difficult to explain, Graeme Clarke agreed that there was an attitude of 'before and after *Tracy*' that lasted about ten years.

Many of the people who shared with me their memories of that dreadful storm said that their most vivid memory was of the noise of the wind and the banging of debris as *Tracy* tore their house apart. A number of them, including Milton Drew, said that they later chopped down large trees in their yards, because every time the wind blew, the noise whistling through the branches reminded them of *Tracy*. Cherry Perron said that she 'cannot stand a tinkling mobile phone or a wind chime'. There are other reminders. John and Janice Woodcock, who are keen gardeners, said that even today they often come across nails and bits and pieces of debris in their garden beds.

People like Chris Kingston-Lee had good reason to leave town after he lost his house and his livelihood, but most people, like Rick Conlon, felt that their greatest loss was the friends who left Darwin and never returned. Rick explained that the greatest memory he had of Darwin—and of *Tracy*—was the 'great community spirit that existed. The camaraderie. People saw mates they thought had gone and openly hugged each other in relief.' Peter Coombe regretted the effect of the stress that the storm placed on some of his friends. Some relationships and marriages were found wanting under the pressure, he said. Judy and Tony Pickering thought that 'the strong feeling of people helping each other and supporting each other was great. It bonded the survivors together and reduced the impact of the disaster.'

Geoff and Barbara Crane lost all of their family photographs. They described feeling like their life before *Tracy* didn't exist, because all of their reminders of that time have gone.

Some people took longer than others to recover from the storm, and this placed stress on some relationships. Lesley Mance's brother John left the NT Police because his wife Rita couldn't bear to stay in Darwin. Elspeth Harvey thought the biggest loss was that of an era; the population changed and the town grew to such an extent that many of the people in town were strangers. Elspeth loves the town, but cannot bring herself to visit the Tropical Cyclone *Tracy* exhibit at the Northern Territory Museum and listen to the recorded noise of the storm. Ann Gray cannot stand the noise of tin scraping or a strong wind, but still won't leave her beloved Darwin. Her greatest loss was her wedding album. She recalled visiting a fortune teller while on holiday in Japan in 1973; she was told that she was in danger and 'should go south'. Twenty-seven years later, she is still in Darwin.

Alan Grove believes the greatest fear he experienced in the storm was that of the unknown. He didn't know how long it would last or how strong it would be, or whether the next gust would be the one that took their lives.

But *Tracy* did bring with it opportunities to rebuild and, for those willing to make some sacrifices, a chance to make money as Darwin rose again from the wreckage.

Laurie Gwynne had very good reasons not to return to Darwin, but he did. He admits today that he didn't want the storm 'to beat' him. He added that he thought he had lost faith in his ability to look after his family, and yet his bravery during the second wind cannot be disputed. Shirley Gwynne's greatest fear was that she would be the only one left alive in the morning. She recalled living in a caravan (something she swears she will never do again) for almost a year afterwards, and constantly stubbing her toes in the confined space. Both Laurie and Shirley remember their neighbours who were killed in Wagaman Terrace behind their house. For Shirley, like the other survivors I spoke with, the material things in life no longer matter. 'They are nice, but they're not the most important things in life.' Tricia Clarke said that she was now 'just as happy to drink out of a Vegemite jar as Waterford crystal . . . People's values changed.'

Tony Pickering stayed in Darwin because he had business interests there that would recover in the long term. He added, '*Tracy* was just piss and wind, and I didn't want to be defeated by it.' Judy

Pickering recalled that they must be stubborn, because twenty-one years after they rebuilt their house it burned down, leaving them with just a kitchen table. Many people who lost everything in the cyclone were philosophical because 'they still had their lives,' she said. John Ryan commented, 'Everyone was in the same boat, so how do you compare? It makes you appreciate what you have in life.' But his wife Sue said she could never go back. 'I couldn't go back to that scene of terror.'

Many people cannot bear to hear the siren noise that accompanies the cyclone alerts, as it reminds them of *Tracy*. Barbara Huddy, especially, gets shivers down her spine when she hears the sound, and it still scares her daughter Nicole. Despite this, the Huddys feel that they gained 'something out of the cyclone' and are better off for having had the experience. Greg commented that that when they talk about the storm today, 'We still get goosebumps.'

After *Tracy*, the attitude of many people that 'Life's too short' may have accounted for the many marriage break-ups that followed, Julie Tammo believes. She had a nervous breakdown almost three months after the storm. 'It affected our relationship,' she said. 'It was hard, but Heikki and I are still together and we worked it all out. I sometimes have sleep problems and I'm terrified of storms.'

Grant Tambling believes that the disaster accelerated the move to self-government in the Northern Territory. He thinks the social reconstruction of the town was difficult owing to the competing interests in the population. Community organisations that had flourished in Darwin took five to seven years to regroup and start up again, he said. Marshall Perron commented, 'Lots of dead wood was blown away, and this allowed a fresh start.' It was estimated in the mid-1990s that 70 per cent of Darwin's population were people who had settled there since the cyclone. John Woodcock said, 'Darwin was changed—for the better. In reality, the town was old and run-down. Now we have a nice city with good, well-built houses. It's a much better city and it's still a nice town. *Tracy* changed the culture; the city grew, we got self-government, and Darwin was no longer a revolving door of public servants.'

Darwin today is a thriving, bustling town with a great emphasis on tourism, especially during the dry season when thousands make

A memorial to those who perished in Cyclone *Tracy*, constructed of six twisted telephone poles—a stark reminder of the sheer power of the storm. (PHOTO MCKAY COLLECTION)

their way north to enjoy Kakadu and the tropical lifestyle. Darwin may have lost its small town friendliness, but not its true character.

Tropical Cyclone *Tracy* might have blown Darwin away, but she didn't blow away the character or the determination of its citizens, who rebuilt their town and started over again. The storm brought a new vitality, and eventually self-government, to the Northern Territory. It brought out the best in many and the worst in a few others; above all, it made people aware of their own mortality. As a reminder of the awesome power that Tropical Cyclone *Tracy* wielded on Christmas Day, 1974, a memorial has been erected near the Casuarina High School comprising six twisted telephone poles. Set in black concrete, it forms a stark reminder that nothing on this planet can be as powerful as Mother Nature.

APPENDIX: THE HEART OF A TROPICAL CYCLONE

Air flows out at top of cyclone—it flows out
very fast, letting more air come in below

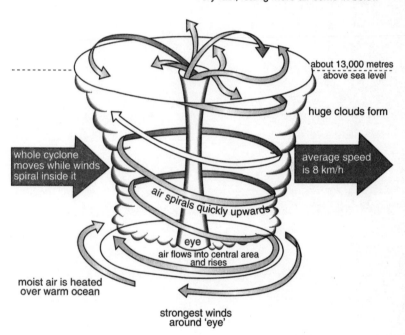

about 13,000 metres
above sea level

huge clouds form

whole cyclone
moves while winds
spiral inside it

average speed
is 8 km/h

air spirals quickly upwards

eye
air flows into central area
and rises

moist air is heated
over warm ocean

strongest winds
around 'eye'

Source: Adapted from S.K. Bourke, *A Geographer's World*, Figure 3.14

ACKNOWLEDGMENTS

This book could only have been made possible through the cooperation and support of many people. I would like to thank my publisher, Ian Bowring, for suggesting the project and for his unswerving support. Thanks go also to my editors, Colette Vella and Robyn Flemming, for their work on the manuscript.

I thank most deeply and sincerely my wife Gay for her support as I continue to pursue my career as a writer and oral historian. I also thank Judy and Tony Pickering and Mark and Debbie Leahy, who helped me enormously and most generously while I was researching and interviewing in Darwin. My appreciation to meteorologist Geoff Crane, who explained to me what a tropical revolving storm is and that forecasting is more than just a matter of looking out a window, and to geographer Malcolm Stacey for his patience in answering my interminable questions.

I am indebted to the staff of the Northern Territory Archives Service for allowing me to use interview transcripts from Laurie Coffey. The use of excerpts from the *Navy Newspaper* is also gratefully acknowledged.

Finally, without those survivors of Cyclone *Tracy* who bravely and generously shared with me their memories of that terrible night, or those others who provided additional advice and support, this book would not have been possible. I extend my heartfelt thanks to:

Carl Allridge, Adelaide
David and Judy Allsop, Darwin
Ronald Assenheim, Bundaberg

John and Helen Auld, Adelaide
Fiona Bekker, Adelaide
Ian and Jane Bowring, Sydney
Rowan and Marcia Charrington, Darwin
Graeme and Tricia Clarke, Darwin
Chris Cleveland, Broome
Matthew Coffey, Darwin
Chris Collins, Darwin
Peter Collins, RAN
Rick Conlon, Darwin
Peter Coombe, Adelaide
Christine Cox, Northern Territory
Cullen Bay Apartments (staff), Darwin
Geoff and Barbara Crane, Brisbane
Kay Dabovitch, Perth
Darwin Port Authority staff
Milton Drew, Brisbane
Lesley Duff (nee Mance), Brisbane
Dianne Ferguson, Brisbane
Josephine Foreman, Brisbane
Francis Good, Northern Territory Archives
Bill Gough, Mt Isa
Peter and Ann Gray, Darwin
Philip Grice, Darwin
Alan Grove, Darwin
Laurie and Shirley Gwynne, Brisbane
Michael and Elsbeth Hannon, Darwin
Elspeth Harvey, Darwin
Klaus Helms, Gove
Greg and Barbara Huddy, Brisbane
Werner Jamnik, Darwin
Kevin and Karen Jurek, Brisbane
Peter and Diane Kerntke, Brisbane
Chris Kingston-Lee, Brisbane
Mark and Debbie Leahy, Darwin
Jenny Lonergan, Brisbane
Fotis Loukaras, Brisbane

Peter McIver, Darwin
Michael McKenzie, ABC Radio, Darwin
Mitchell Library Archival staff, Sydney
Elizabeth Morris, Deputy Coroner, Darwin
Mick Murdoch, ABC Radio, Darwin
Museum of Northern Territory (Sue Harlow and Mickey Dewar)
Northern Territory Archives staff
Marilyn and Bob Opperman, New South Wales
Max and Rhonda Ortmann, Darwin
Nick Paspaley, Darwin
Rob Perkins, Brisbane
Marshall and Cherry Perron, Darwin
Jude and Tony Pickering, Darwin
Eddie and Greta Quong, Darwin
John and Sue Ryan, Adelaide
Malcolm Stacey, Parkes
Grant Tambling, Darwin
Julie Tammo, Darwin
Nat Tunley (nee Guinane), Brisbane
Trevor Wardrope, Police Sergeant, Coroner's Office, Darwin
Judith Watson, Brisbane
John and Janice Woodcock, Darwin

INDEX